THE END OF SUSTAINABILITY

ENVIRONMENT AND SOCIETY

KIMBERLY K. SMITH, EDITOR

The End of Sustainability

Resilience and the Future of
Environmental Governance in
the Anthropocene

Melinda Harm Benson
and
Robin Kundis Craig

Published by the University Press of Kansas (Lawrence, Kansas 66045), which was organized by the Kansas Board of Regents and is operated and funded by Emporia State University, Fort Hays State University, Kansas State University, Pittsburg State University, the University of Kansas, and Wichita State University.

Library of Congress Cataloging-in-Publication Data available at

Names: Benson, Melinda Harm, author. | Craig, Robin Kundis, author.
Title: The end of sustainability : resilience and the future of environmental governance in the Anthropocene / Melinda Harm Benson, Robin Kundis Craig.
Description: Lawrence, Kansas : University Press of Kansas, 2017. | Series: Environment and society | Includes index.
Identifiers: LCCN 2017038270
ISBN 9780700625161 (hardback)
ISBN 9780700625178 (ebook)
Subjects: LCSH: Environmental management. | Sustainable development. |Geology, Stratigraphic—Anthropocene. | BISAC: NATURE / Ecology. | LAW / Environmental. | POLITICAL SCIENCE / Public Policy / Environmental Policy.
Classification: LCC GE300 .B463 2017 | DDC 304.2—dc23
LC record available at https://lccn.loc.gov/2017038270.

British Library Cataloguing-in-Publication Data is available.

Printed in the United States of America

10 9 8 7 6 5 4 3 2 1

The paper used in this publication is recycled and contains 30 percent postconsumer waste. It is acid free and meets the minimum requirements of the American National Standard for Permanence of Paper for Printed Library Materials Z39.48-1992.

Contents

Preface

This book is the result of our discussions and collaborations over the past several years. We first met in 2010 and both attended several academic conferences where "resilience" was becoming the *mot du jour*. One of the first conversations the two of us had was about whether "resilience" and "sustainability" were actually different concepts and whether resilience was simply a new buzzword describing old ideas. We were skeptical, but we were also frustrated with the status quo. At the time, hope was fading regarding President Obama's commitment to addressing climate change. When he came into office in 2009, he promised that health care and climate change would be the top two issues on his agenda. Two years into his presidency, it was obvious that he only had the political capital for one of those challenges—and he chose health care. Climate change policy was at a standstill internationally, as well. Negotiations at the United Nations Climate Change Conference in Copenhagen in December 2009 failed to produce a new climate agreement and were considered a dismal failure.

It seemed to us that the rhetoric surrounding climate change was business as usual: sustainability is the goal, development is the path, and we can have it all. In 2010, Robin published a groundbreaking article in *Harvard Environmental Law Review* titled "'Stationarity Is Dead'—Long Live Transformation."[1] In it, she encouraged policymakers to acknowledge what scientists had already accepted: that stationarity (the idea that natural systems fluctuate only within an unchanging envelope of variability) is dead. The end of stationarity means that global climate change and other factors related to human modification of the Earth are quickly eliminating our ability to rely on past assumptions about ecological systems, including the pace and scale at which they change. When Melinda checked in with Robin a year later to see if the article had gained much attention, Robin reported she'd received little outside of academia, which was frustrating to us both. We hatched a plan to collaborate on a more provocative piece— one that would gain the attention of a broader audience and generate discussions around climate change and the challenges to come. By that time, both of us had come to the conclusion that resilience actually is (or at least has the *potential* to be) more than a buzzword. We wrote a law review article on "Replacing Sustainability" for a special symposium edition of the *Akron Law Review* on the next generation of environmental law, and we published the essay "The End of Sustainability" in *Sustainability and Natural Resources.*[2]

Both of these articles began attracting significant attention from academics and various forms of social media. We continued to further our ideas in *Ensia* magazine,[3] a webinar sponsored by the Environmental Law Institute, and at various symposia and conferences, including the University of Missouri School of Law's 2014 symposium, "Environmental Law 4.0: Adaptive and Resilient," sponsored by the *Journal of Environmental and Sustainability Law* and the University of Arizona College of Law's annual conferences on sustainability. Many of the ideas found in this book were first published in other volumes and are cited throughout the manuscript.

In 2014, we were contacted by Kim Smith, a professor at Carlton College and editor of the Environment and Society series for the University Press of Kansas. She convinced us that (having convinced *her*) our message on the end of sustainability deserved a book-length manuscript. And so here it is, almost three years later.

The message that stationarity is dead and sustainability is no longer an option is not a popular one. We find this reaction particularly true among those who have dedicated their careers and scholarship to the promotion of sustainability and sustainable development. Over the past few years, we've been described as everything from naïve to nihilistic. One online commentator on our *Ensia* piece, for example, stated that "an embrace of an 'Anthropocene' of unknown dystopic dimension is rank foolishness."

For many, sustainability is such a worthy ideal that they refuse to give it up. We ask: do we really have a choice? Far from being naïve, we are convinced that most of those who disagree with us simply do not comprehend the enormity of climate change and the transformations it will bring. This book is our attempt to remedy that. Far from being nihilistic, we believe that past narratives about climate change are either unrealistic or disempowering. We propose a new narrative and accompanying communitarian approach to environmental governance that integrates humans back into the system—a complex social and ecological system known as Earth. This is a world characterized by radical uncertainty and unprecedented rates of change, and yes, probably great loss, but one in which humans and natural systems—in short, the greater community of life on Planet Earth—remain capable of adaptation and successful transformation.

Chapter 1, "Welcome to the Anthropocene," explains why we agree with recent recommendations to denote our current epoch as the Anthropocene—a new era in which humans are the key agent of change on the planet. We then make the case that we need a new approach to environmental governance in the Anthropocene, one informed by a new narrative based on understanding and responding to climate change. The chapter describes in sobering detail the challenges that global climate change will bring and highlights past narratives about climate change and why they won't work moving forward.

Chapter 2, "Narrating Our Relationship with Nature," explores environmental and natural resources law and policy and the underlying cultural narratives on which they are based. Focusing on the United States, these stories include the pre–environmental awareness narrative of Manifest Destiny; what we call the "tragedy" narrative of the early environmental movement in the United States in the 1970s; and the sustainability narrative. A nation's laws and policies are inevitably bound up in the dominant cultural realities of that country, making it tricky to generalize from one legal system to another. For this reason, while we expect that many of the discussions in this book will resonate in other countries and cultures, we consciously limit its focus to the United States, arguing that this country in particular needs new cultural and legal narratives to successfully cope with the Anthropocene.

In Chapter 3, "Resilience and the Trickster: A New Narrative for the Anthropocene," we more fully introduce the concept of resilience and resilience theory and offer it as a new narrative for the Anthropocene. Our articulation of this narrative has both an ecological and a cultural component. It embraces system theory and key concepts from the C. S. "Buzz" Holling school of social-ecological resilience, which describes social-ecological systems (SESs) as always being in continual flux, subject to complex, multi-scalar interacting drivers and feedbacks. A key reason why resilience theory is an appropriate conceptual framework for the Anthropocene is its recognition of "surprise" and acknowledgment of the inherently unpredictable nature of SESs.

We also argue that, at least in the United States, resilience theory currently lacks a corresponding cultural narrative, impeding its ready incorporation into social and political products like law. Resilience theory itself does, however, point in the right direction. Its concept of "panarchy," a term coined by Lance Gunderson and Buzz Holling, is named after the Greek trickster god Pan to invoke capricious ways of being. For resilience to become the right cultural narrative for the Anthropocene, Americans must embrace this embedded trickster narrative of challenges and change. The trickster figure is notably absent from European-based American culture but present in many others, notably Native American cultures. In this chapter, we discuss the power of the trickster and frame climate change as one manifestation of the trickster: an agent of change.

The next two chapters provide case studies that illustrate the concepts explored in the previous chapters. We intentionally chose two very different systems operating at radically different scales. Chapter 4, "Regime Change for New Mexico Watersheds," describes an SES in a state of transition. A combination of climate change, bark beetle infestation, and wildfire is shifting New Mexico's forest watersheds in irreversible ways. Efforts to understand the implications of crossing this ecological threshold and its impacts on downstream water supply provide a clear example of how adaptive capacity is needed within the social system in order to respond to the challenges that will be associated with climate change. This case study presents a dramatic ecological challenge and some encouraging social responses.

Chapter 5, "Marine Fisheries and Biodiversity: How the Trickster Undermines Sustainable Yield," examines ocean governance—specifically, the mismatch between current fisheries management and the ecological realities of the Anthropocene. Fisheries management evolved from the paradigm of inexhaustibility—a belief that humans were too puny to ever significantly impair the ocean's vast resources—into management that pursues maximum sustainable yield. However, the rapidly accelerating and synergistic changes in the ocean increasingly threaten marine biodiversity. Because many drivers of marine changes are beyond immediate human management, however, this chapter focuses instead on how incorporation of resilience theory could and should change fisheries management.

Chapter 6, "Thinking Like a System: Resilience as a Narrative of Connection," invokes Aldo Leopold's iconic concept of "thinking like a mountain" to call for communitarianism as a critical component of governance and resilience theory moving forward. We explain that Leopold was a resilience scholar before the concept even existed, and his key message that humans are a part of a land community was radical both for his time and for today. Leopold challenged the humanist, Enlightenment tradition that separates humans from the natural world and offered a more communitarian approach known as "the land ethic." Taking a more critical look at applications of resilience theory to date, this chapter notes the ways in which this work can be strengthened by following Leopold's lead. It uses the concept of property rights and societal limits on those rights to point out how resilience theory has the potential to take a normative turn and become a narrative of connection.

Our Conclusion, "Living the New Story: Implications for Governance," provides a series of recommendations for how laws and policies can change to reflect a resilience-based approach to the Anthropocene. It adopts and expands the concept of "principled flexibility" that Robin first articulated in "'Stationarity Is Dead,'" focusing now on what normative principles a resilience theory framework might suggest for the law and policy of the Anthropocene.

Our hope is that this book generates conversations about how we frame and respond to climate change and the Anthropocene's other challenges. It provides introductory material for those new to the Anthropocene, resilience theory, and environmental governance. It also includes material even seasoned resilience scholars will find new and provocative—particularly our call for a normative turn in resilience scholarship in Chapter 6 and our recommendations for shifts in US law and policy in the Conclusion.

Life in the Anthropocene demands a new approach to environmental governance. This book is our attempt to begin shifting governance paradigms away from sustainability and toward a conceptual framework that embraces continual change. We need new tools and new stories that will enable us to meaningfully respond to what climate change will continue to bring, and we hope that this book serves as the starting point for a new and far more productive conversation on the subject of law and policy for the Anthropocene.

Acknowledgments

We owe a debt of thanks to Kim Smith, as well as to our initial editor Charles Myers and current editor Kim Hogeland for their encouragement and support. We also want to thank all the other editors, conference and symposia organizers, and others who supported the development of these ideas, as well as the dozens of students and colleagues who joined us in discussions on various themes found in the book. It's been a great conversation that we hope will continue. We thank our editorial assistant Laurel Ladwig, who brought not only a keen editorial eye but also a love for the project that greatly enhanced the final manuscript. Thanks also to Sean T. O'Neill and Andrew Steiner, who provided much-needed technical assistance with many of the figures that found their way into this book. Our loving and supportive husbands, Reed Benson and Don Craig, have our sincere thanks and deepest gratitude.

We also thank our detractors and critics, especially the formidable John Dernbach and Fred Cheever, with whom we have had several public and private debates and with whom we remain friends despite significantly different views on how to best approach the Anthropocene. Fred, tragically, died far too young as this book was going to press, doing one of the many things he loved in the environment: whitewater rafting. We acutely feel his loss on many levels—as an intellectual debate partner, as a ceaseless advocate for the United States' ecological future, and as a longtime friend and colleague.

Additionally, the cover art was generously donated by Heli Kankainen, a Finnish American artist who draws both inspiration and strength from the living things around us. In Finnish folklore, the raven is seen as a bird of ill omen, but it is said that below its wing is a feather of fortune. Ravens also have a significant role in the Finnish national epic, *Kalevala*.

CHAPTER ONE

Welcome to the Anthropocene

The world around us is changing. As the Intergovernmental Panel on Climate Change (IPCC) extensively documented in its *Climate Change 2014 Summary for Policymakers*,[1] "human influence on the climate system is clear, and recent anthropogenic emissions of greenhouse gases are the highest in history. Recent climate changes have had widespread impacts on human and natural systems."[2] Ongoing changes include alterations to air temperature and wind currents, ocean temperature and currents, and terrestrial and weather conditions around the world. In turn, these changes impact the ecosystems, that depend upon those global systems as well as the societies that depend upon those ecosystems, including for their livelihoods and well-being.[3] The impacts of climate change underscore that we are living in the Anthropocene, an era in which humans are a fundamental agent of change on our planet.

The realities of the Anthropocene demand a new approach to environmental governance. To date, natural resource environmental law and policy in the United States have not kept pace with our understanding of how natural systems actually work or the nature and scale of climate change. For this and other reasons, it is important to reexamine current and past approaches to environmental management in light of the Anthropocene. The key argument of this book is that the time has come for us to collectively reexamine—and, we argue, ultimately move past—the concept of sustainability as an environmental governance goal.

While definitions of "sustainable" and "sustainability" vary, the core concept in most invocations is some version of living within one's means—that is, not expending or consuming more than can be maintained or replaced. And that focus is important. "Sustainability" becomes an almost meaningless concept unless it is applied to a particular subject: the sustainability of *what*, exactly? Personal financial sustainability, for example, envisions living within a budget and not mounting up debt when the principal cannot be repaid.

The focus is not all that different in the natural resources context, in which goals of sustainability have to take account both the difference between renewable and nonrenewable resources and the interconnectedness of physical, chemical, and biological systems. For example, nonrenewable resources such as oil and natural gas that are mined raise critical questions regarding what we mean by sustainability. Similarly, pumping groundwater from an aquifer that will take millennia to recharge is by definition unsustainable, regardless of how many decades it takes for that point to become clear.

Defining the sustainability of renewable resources also takes a bit of thought. For example, a "sustainable" harvest of timber might mean taking no more trees than can be replaced before the next harvest. However, that rather simplistic approach ignores the fact that trees play many roles. Ideally, the calculation of a "sustainable" harvest would also take account of the trees' role in providing a forest for recreation, the other forms of life that depend on those trees for shelter and for food, the role of the trees in soil production and retention, the trees' contribution to water storage and release, their production of oxygen and uptake of carbon dioxide, and so on.

Done right, pursuit of "sustainable" use of natural resources requires a fairly elaborate analysis of what those natural resources are actually doing in and across several scales of complex and changing social-ecological systems (SESs) as well as an acknowledgement that any change in resource use will almost always come with trade-offs, however minor they may be: a tree used to build a house is the loss of a tree that could sequester carbon dioxide. More fundamentally problematic, however, is the fact that the pursuit of sustainability in natural resources governance has inherently assumed a certain level of predictability and stationarity in these systems—that the world tomorrow will behave as the world does today *despite* human intervention. As Chapter 3 will explore in detail, this assumption ignores the core dynamism of ecological systems and hence has long been in need of refinement.

Unpredictability and change undermine sustainability goals. To go back to the personal finances example, significant and unpredictable fluctuations in income from week to week, month to month, or year to year make it difficult to establish a sustainable budget. Similarly, the unpredictability of the relevant national and local economies might be critical factors in how large the budget should be, such as during periods of high inflation rates. Prudent households facing these kinds of unpredictable financial realities are likely to take a precautionary approach to spending and, if possible, establish a substantial savings account to bridge large variances in income and spending power.

Today, natural resources management, law, and policy face similar dilemmas as a result of change and unpredictability. In the Anthropocene, the simplistic approach to natural resources sustainability displayed in the timber harvest example ignores the realities of climate change and other anthropogenic influences on complex SESs. The effects can be dramatic: as climate-related pine beetle infestations and forest destruction in Colorado and British Columbia attest, we might not in fact get a next harvest because the underlying ecological support systems for tree growth are themselves changing.

The continued invocation of sustainability in international talks, development goals, and other policy discussions ignores the emerging reality of climate change and its implications for our ability to define—let alone achieve—sustainability as the natural world is altering under our feet. It's not that sustainability is not a laudable ideal; the issue is whether the sustainability narrative is still a helpful way of conceptualizing environmental governance goals. Two major elements of the sustainability story are clearly worth transmitting into the Anthropocene: the idea that we cannot consider environmental, economic, and social issues in isolation and the importance of *inter-* and *intra-*generational equity. However, the Anthropocene demands difficult conversations about the trade-offs among economic development, social improvement, and environmental protection, as well as the future costs of present actions and the inevitable inequities that will result.

In this introduction, we present the nature of the Anthropocene. We then provide a brief outline of the environmental challenges that climate change will bring in this new era and discuss the cultural narratives that US society has used to date to understand climate change. We argue that these current narratives have disempowered us and reduced our capacity to address climate change. More broadly, however, climate change narratives are a specific manifestation of the more general and increasing gap between ecological understandings of complex natural systems and environmental and natural resources law and policy. We explore this more general historical evolution in Chapter 2, before proposing in Chapter 3 that the narrative of resilience is a more productive new way of conceptualizing environmental and natural resource management in the Anthropocene. Resilience is not itself an environmental goal. Instead, resilience theory is a theoretical framework for conceptualizing how complex SESs respond to continual change and how those responses might affect human environmental management efforts.

While not inherently incompatible concepts, resilience and sustainability are not the same. The pursuit of sustainability inherently assumes that we: (1) know what can be sustained and how; and (2) have the capacity to maintain some type of stationarity and/or equilibrium in the relevant systems. These assumptions are not appropriate in the Anthropocene (and they probably never were). By contrast, resilience theory does not assume stationarity within ecological and planetary systems and instead acknowledges disequilibrium and nonlinear change in SESs. It embraces the dynamics and complexities of SESs, does not require certainty, and emphasizes building adaptive capacity rather than maintaining stationarity. For these reasons among others, resilience theory provides a more realistic approach to environmental governance in the Anthropocene.

What Exactly *Is* the Anthropocene?

The term "Anthropocene" acknowledges that human action has become an important driver—arguably *the* most important driver—of ecological change. It recognizes that humans are pervasively altering planetary ecological function.[4] The most authoritative definitions of "Anthropocene" come from the International Union of Geological Sciences, the international scientific organization that is in charge of officially designating and naming geological time periods, and the International Commission on Stratigraphy (ICS), which evaluates the scientific evidence in support of new geological period designations. On August 29, 2016, the ICS's Working Group on the Anthropocene recommended that our current interval be recognized as a new epoch, the Anthropocene,[5] or "the new age of humans."[6] It noted that:

> Paul Crutzen and Eugene Stoermer [coined the term "Anthropocene"] in 2000 to denote the present time interval, in which many geologically

significant conditions and processes are profoundly altered by human activities. These include changes in: erosion and sediment transport associated with a variety of anthropogenic processes, including colonisation, agriculture, urbanisation and global warming[,] the chemical composition of the atmosphere, oceans and soils, with significant anthropogenic perturbations of the cycles of elements such as carbon, nitrogen, phosphorus and various metals[,] environmental conditions generated by these perturbations; these include global warming, ocean acidification and spreading oceanic "dead zones"[,] the biosphere both on land and in the sea, as a result of habitat loss, predation, species invasions and the physical and chemical changes noted above.[7]

While these alterations to natural systems derive from many causes—consumption of natural resources, pollution and waste disposal, and a growing human population—the Working Group is currently dating the Anthropocene to about 1950, reflecting the dispersal of radioactive materials from nuclear bomb tests across the planet. Another possibility for the starting point, however, is the Industrial Revolution, underscoring the importance of climate change as the most recently acknowledged, pervasive, and complex of these human drivers of ecological change.

Beyond strict geological conventions, the concept of Anthropocene is already being extensively used in a variety of academic and policy contexts.[8] At this point, however, the Anthropocene and its implications are not reflected in environmental and natural resources law in the United States. As we shall see, current approaches tend to view humans as separate from the environment, most often as agents who act as the controlling engineers who can shape environmental processes to our own desires and goals. The notion of an Anthropocene provides a potentially powerful tool for reconceptualizing our relationship to the natural environment. Specifically, the concept of the Anthropocene underscores the fact that humans are one component of vast networks of complex SESs—systems that sometimes react in ways that humans did not intend or, often, could not have even predicted. These systems operate in multiple temporal and spatial scales, some of them fast and local, others plodding and regional, and still others millennial and global.

As an example, the immediate and local environmental impact of the Industrial Revolution was air pollution created by burning coal, producing the smog and the "killer fogs" in London and many US industrial cities that occurred when inversion layers in the atmosphere trapped soot and other air pollutants close to the ground. There were no environmental laws in place at the time to address these issues because the Industrial Revolution brought with it radical changes to our use of fossil fuels, creating new pollution problems beyond anything previously imagined. Nevertheless, these events did lead to new laws to address air pollution—in the United States, the Clean Air Act, enacted in its current structure in 1970.

However, the air pollutant products of the Industrial Revolution, particularly carbon dioxide, also impact medium- and large-scale planetary systems, and we are just now beginning to feel the effects of the Industrial Revolution at these scales. At the medium time scale is climate change, resulting from the increasing concentration of carbon dioxide in the atmosphere over a few centuries, with planet-scale implications. At the large scale is ocean acidification, the progressive lowering of the ocean's pH as it absorbs excess carbon dioxide from the atmosphere to cycle it, eventually, back into the earth's rocks and crust. Like climate change, ocean acidification is planetary in geographic scale, but scientists estimate that it will take at least a thousand years to cycle carbon dioxide back out of ocean waters, meaning that ocean acidification reflects system processes that operate on a temporal scale at least an order of magnitude greater than climate change.

Humans did not *intend* to cause air pollution, climate change, or ocean acidification as we increasingly exploited fossil fuels—but that's the point of the Anthropocene. Humans can affect, and in turn be affected by, larger-scale planetary systems in ways that we neither intended nor predicted. We are embedded within these complex earth systems at all scales, clearly able to influence them but also increasingly clearly not in complete control of them. Humans are both part of something bigger than our species and subject to system dynamics that we cannot completely master.

Perhaps unsurprisingly, responses to the Anthropocene often reveal anxiety and feelings of disempowerment. Among ecologists and biologists, for example, Paul Robbins and Sarah Moore have identified two conflicting reactions to the Anthropocene in the context of ecological restoration efforts. "First is the clear and abiding concern—or obsession—with human transformation of the earth to a point of irreversibility," a reactive narrative that sounds in tragedy, value judgments, and an urgent call to restore ecosystems to prehuman status.[9] Second is a deepened concern regarding the possibility of scientific objectivity, a recognition that "in a quickly transforming environment, deeply held human biases (like those towards nativeness) cause apparently scientific assessments of change to be fraught with normative assumptions."[10] Perversely, these two camps figure human influence as either obsessively to be eradicated or obsessively inescapable. As Robbins and Moore themselves summarize, "At precisely the emancipatory moment that ecological science has transcended the flawed expectation that a single ecological condition can provide the blueprint to regulate and guide human behavior—whether nature, wilderness, or the biogeography of the pre-Columbian period—the community ironically finds itself paralyzed by acknowledgment of human agency on the earth and the normative character of science itself."[11]

Instead of producing anxiety and fear, the Anthropocene should encourage a new sense of humility and respect for natural systems and our role in them. In turn, this new perspective should be reflected in environmental and natural resources law and policy. Law and policy face an increasing need to acknowledge the reality that the natural world operates in complex systems subject to continual change. *The Anthropocene offers an opportunity to realign law and policy with ecological realities.* Climate change and related factors are changing the Earth in dramatic and fundamental ways. We must embrace new orientations that help us cope with a world characterized by vast uncertainty, radical complexity, and continual change. Our understanding of this new ecological reality and our assumptions about its causes and effects should be reflected in our laws and policies, and the new geological epoch should be accompanied by a new orientation toward climate change and other global environmental challenges. The core arguments in this book are that: (1) the Anthropocene requires us to reframe how we think about climate change and other environmental challenges; and (2) "sustainability" is no longer a viable paradigm.

Here, we outline some of the impacts climate change will have on our capacity to formulate narratives. We then discuss the current narratives about climate change specifically as an introduction both to the importance of cultural narratives and to the disempowerment that arises when cultural narratives do not match ecological realities.

Climate Change and Cultural Narratives
Many human-induced changes—habitat destruction, pollution, overuse of natural resources such as fresh water, fisheries, and forests—contribute to the ecological reality of the Anthropocene. Climate change exacerbates all of these challenges because it interacts synergistically with other hu-

man impacts in ways that create unpredictable SES impacts at all temporal and spatial scales. Climate change undermines our ability to predict our future—and, as such, it highlights the fact that environmental cultural narratives in the United States are not promoting our ability to cope with the Anthropocene.

Cultural narratives are stories told at the societal level, deeply embedded stories that frame and contextualize events within a particular culture to help give them meaning. They include but are not limited to creation stories or other stories about the community's origins; fables or other stories that teach norms of moral, ethical, and social values and behavior; and cultural histories or stories of self-identification and description.[12]

Many of these narratives conceptualize both how change occurs and how humans relate to their environment. As will be discussed in more detail below, for example, the United States has a particularly strong cultural narrative of apocalypse that resonates both religiously and secularly, making it particularly easy for Americans to narrate climate change as the end of the world.

More importantly, our cultural narratives of change—what might be termed the cultural psychology of change—influence how we actually deal with ecological change. Research demonstrates that we humans actually understand our world through story. We are narrative beings. Indeed, humans are inclined to see narratives even where there are none. For example, in a 1944 study, Fritz Heider and Marianne Simmel showed students a short film of geometric shapes. When asked to explain what they had seen, all but one of the students spontaneously created a narrative that anthropomorphized the shapes.[13] Narrative also appears to be a necessary neurological component of certain human behaviors, such as cooperation—not a trivial finding for the Anthropocene, where solutions to complexly "wicked" problems like climate change require extensive cooperation.[14]

Moreover, the stories we tell ourselves about what is happening to us can literally change our ability to respond. In *Words Can Change Your Brain*, psychologists Mark Waldman and Andrew Newberg summarize their research using fMRI scanners to examine neural changes happening in the human brain via dozens of stress-producing hormones and neurotransmitters as they react to both negative thoughts and words such as "no" and to positive thoughts and words. The results reveal that the stress responses induced by negative thoughts and emotions immediately interrupt the normal functioning of the human brain by impairing logic and reason.[15] Repeatedly telling ourselves that climate change is the apocalypse, in other words, may actually impede our ability to respond productively.

Thus, among other things, cultural narratives are important because they both reflect and influence individual psychology: The stories we tell at an individual level tend to reflect the cultural narratives in which we are embedded. Changing the relevant cultural narrative, therefore, can be a very powerful way to change both social phenomena like law and individual behavior. Cultural narratives that no longer reflect a changing reality, in contrast, can prompt both social and individual behaviors that are maladaptive.

And our reality *is* changing, as climate change underscores. The IPCC's 2014 reports make clear that climate change increases the risks that ecosystems and SESs will cross thresholds (that is, boundary conditions whose crossing signals or promotes a transformation in the system) more frequently and in unpredictable and often undesirable ways. Thus, "the precise levels of climate change sufficient to trigger abrupt and irreversible change remain uncertain, but the risk associated with crossing such thresholds increases with rising temperature (*medium confidence*)."[16] More specifically, "magnitudes and rates of climate change associated with medium- to high-emission scenarios pose an increased risk of abrupt and irreversible regional-scale change in the composition, structure and function of marine, terrestrial and freshwater ecosystems, including wetlands (*medium confidence*). A reduction in permafrost extent is virtually certain with continued rise in global temperatures."[17]

Even if we manage to stabilize greenhouse gas concentrations in the atmosphere, the complex and multi-scalar planetwide systems of which we are a part cannot respond immediately. For this reason, we will be dealing with a changing climate, developing impacts, and increasing risks of threshold crossings for centuries.[18] For example, the IPCC reported that "shifting biomes, soil carbon, ice sheets, ocean temperatures and associated sea-level rise all have their own intrinsic long timescales which will result in changes lasting hundreds to thousands of years after global surface temperature is stabilized."[19] "There is high confidence that ocean acidification will increase for centuries if carbon dioxide emissions continue, and will strongly affect marine ecosystems," and "it is virtually certain that global mean sea-level rise will continue for many centuries beyond 2100, with the amount of rise dependent on future emissions."[20]

Thus, *continual social-ecological change is our new normal for the fore-*

seeable future. Notably, because the IPCC reports represent the consensus view of the world's leading climate scientists, the IPCC's projections tend to be conservative. Indeed, when new science emerges that suggests that the IPCC's projections need correction, it's often because climate change impacts—Antarctic ice sheet destabilization, sea-level rise, coral reef bleaching, species extinction—are occurring faster and more dramatically than initially anticipated. However, even the conservative IPCC now concludes that the Earth is experiencing unprecedented alterations as a result of a changing climate, including impacts on both natural ecosystems and human societies. Further, it now emphasizes that "continued emission of greenhouse gases will cause further warming and long-lasting changes in all components of the climate system, increasing the likelihood of severe, pervasive and irreversible impacts for people and ecosystems."[21]

Humans can, as the IPCC emphasizes, still reduce the magnitude of climate change impacts through aggressive climate change *mitigation* efforts—that is, policies and actions to reduce greenhouse gas emissions and (possibly) technological innovations to remove existing greenhouse gases from the atmosphere.[22] Indeed, failure to mitigate just makes it more likely that widespread and irreversible changes to ecosystems and SESs will occur.[23]

Nevertheless, while we acknowledge that the world should be pursuing multiple effective strategies to aggressively reduce greenhouse gas emissions, this book focuses instead on the necessity of *adaptation*—societal efforts to cope with the changes that are occurring and will continue to occur in ecosystems and planetary processes and the human societies embedded within both. Even aggressive mitigation cannot stave off continual change for the foreseeable future, and our cultural narratives need to be able to handle this new reality. Specifically, we take to heart the IPCC's recognition in 2014 that "adaptation planning and implementation at all levels of governance are contingent on societal values, objectives and risk perceptions."[24] In other words, our ability to effectively adapt to climate change depends in part on the realities of particular cultures and the stories that we tell ourselves, our cultural narratives.

Emerging Climate Change Narratives in the United States to Date

Cultural narratives mediate how we both respond to and understand change.[25] Unfortunately, the dominant US culture currently lacks a narrative about climate change that is both empowering and realistic. We need a narrative that enables us to live with the Anthropocene. Specifically, we posit that Americans desperately need a new cultural narrative that allows us to productively cope with continual and increasingly unpredictable change in both our natural and our social systems. So far, we have failed to produce an empowering new cultural narrative with which to navigate these changes.

By underscoring our new reality of continual and unpredictable change, climate change *should* be fundamentally challenging Americans' ability to effectively narrate, and hence effectively engage, our evolving role in SESs. The concept of the Anthropocene teaches us that humans are an influential but not determinative component of some very complex SESs. The anthropocentrism of climate change is the quintessential example of this influence. It should dismiss, once and for all, any assumption that humans are separate from nature. It should also invoke profound humility in the face of the unintended consequences of human development.

Instead, popular culture has embraced polarizing, paralyzing, and disempowering narratives about climate change. As Andrew J. Hoffman argued in *How Culture Shapes the Climate Change Debate,* in the United States, "climate change has been transformed into a rhetorical contest more akin to the spectacle of a sports match, pitting one side against the other with the goal of victory through the cynical use of politics, fear, distrust, and intolerance."[26] As a result, the stories about climate change that have emerged here either unhelpfully assert humans' technological dominance over nature or, perversely, create tales of human impotence. Four such narratives currently infuse US culture: (1) climate change doesn't really exist; (2) climate change may exist, but humans didn't cause it; (3) climate change exists, but we can engineer our way out of it and its effects; and (4) climate change exists and our current way of life is doomed (with several variations).

Climate Change Doesn't Really Exist

In early August 2007, just after the US Supreme Court decided that the EPA can regulate carbon dioxide and other greenhouse gases under the federal Clean Air Act, *Newsweek* published a highly misleading cover announcing that "Global Warming Is a Hoax." (A footnote made clear that the issue contained a story about climate change deniers, not that *Newsweek* was asserting its own position.) Ten years later, the American president Donald Trump has yet to disavow his long history of referring to climate change as a "hoax," including a November 2012 tweet (which he now claims was a joke) that the Chinese invented climate change to disadvantage American manufacturing. Climate change denial is alive and well in the United States.

As a continuing cultural narrative, the assertion that climate change is not happening is the most disempowering of America's climate change cultural narratives. If the story is that climate change isn't happening, there's no need for any kind of fundamental adjustment to society. By extension, nor is there need to improve US environmental and natural resources law and policy. To state the obvious, the climate change denial narrative promotes continued inaction in the face of climate change, an adherence to a "business as usual" scenario that climate scientists generally agree portends one of the most dangerously altered futures for Planet Earth and its inhabitants.

While the full-on denier narrative in the United States is shrinking overall, it still exists. In 2012, Yale University's "Six Americas" climate change project found that 8 percent of Americans are still "dismissive" of climate change, while another 13 percent doubt that it is occurring and 9 percent are disengaged from climate change issues,[27] suggesting that about 30 percent of Americans effectively subscribe to some form of climate change denial. Indeed, a 2013 survey by the Yale Project on Climate Change Communication concluded that 37 percent of people in the United States do not believe that climate change is happening.[28] In addition, Americans are peculiarly wedded to this narrative compared to the international community: a 2014 survey of twenty countries found that the United States ranked first in climate change denial.[29]

Importantly, the percentage of Americans who deny that climate change is occurring has varied over time. For example, climate change deniers increased from 18 percent in fall 2008 to 48 percent in spring 2010, following the "ClimateGate" e-mail scandal.[30] These observations suggest that changing events can both undermine and strengthen the force of the climate change denial narrative. Nevertheless, its persistence undermines the entire nation's ability to cope with continual, increasing, and increasingly unpredictable change.

It Isn't Us

In addition to being the world's most persistent climate change deniers, Americans also lead the world in asserting that, if climate change is happening, humans didn't cause it.[31] According to Gallup Poll surveys, between 2010 and 2012, only about half of people in the United States believed that humans were causing climate change.[32] While that number increased to 57 percent in 2013 and 2014, about 40 percent of Americans still deny human involvement in climate change.[33]

Like the climate change denial narrative, the "It Isn't Us" cultural narrative is disempowering. Most obviously, this second narrative vitiates any reason to engage in greenhouse gas regulation. If anthropogenic greenhouse gas emissions are not causing climate change, there is no reason for humans to alter our greenhouse gas–producing activities. Again, business and lifestyles can proceed as normal. Thus, by denying the human role in climate change, the "It Isn't Us" narrative effectively undermines any concerted effort to engage in climate change mitigation—that is, in legal efforts to reduce greenhouse gas emissions and, eventually, greenhouse gas concentrations in the atmosphere.

Unlike the climate change denial narrative, however, this second narrative *can* allow for climate change adaptation efforts. Whether those adaptation efforts will become sufficiently urgent, however, is more doubtful, because the "It Isn't Us" narrative comes with two variations: climate change is so natural that it's no big deal, or climate change is a natural disaster on par with earthquakes and hurricanes. Noted scientific climate skeptic Dr. S. Fred Singer has succinctly summed up the first variation of the "It Isn't Us" narrative in interviews: "Climate change is a natural phenomenon. Climate keeps changing all the time. The fact that climate changes is not in itself a threat, because, obviously, in the past human beings have adapted to all kinds of climate changes."[34] While Dr. Singer's faith in human adaptive capacity is in fact empowering, the story that climate change is so natural that we don't need to worry about our adaptation efforts is not. By couching climate change in an aura of "been there, done that" predictability, this variation on the "It Isn't Us" narrative elides the profound surprises that climate change is likely to spring on human societies.

In contrast, viewing climate change as a natural disaster is more likely to spur adaptation efforts. Moreover, if the climate is changing in ways that directly affect human lives, the cause is largely irrelevant to the issue of whether adaptation efforts are necessary. Katrina Kuh has labeled such efforts "agnostic adaptation" and has identified cultural moments when such approaches may be more beneficial than exhortations based on humans being the cause of climate change.[35]

Nevertheless, within this second variation on the "It Isn't Us" narrative, humans become the disempowered victims of malevolent natural forces. As a result, even with respect to climate change adaptation, this narrative thus risks infusing climate change efforts with a profound sense of human limitation: if climate change is just another natural disaster like floods and earthquakes and hurricanes, then there's a limit to what we can do to prepare for and to improve our future.

Technology Will Save Us

In 2014, noting that storm surge and flooding are major climate change risks for coastal cities, researchers publishing in *Science* evaluated a number of adaptation strategies for New York City in light of vulnerabilities revealed by Superstorm Sandy.[36] The scenarios all involved building at least some water-based barriers to flooding. Researchers concluded that, "independent of the discount rate, it is economically effective to invest in a storm surge barrier system in 2040 if climate change develops" in at least a moderately bad way, although the issues involved in building barriers needed to be studied immediately.[37]

While the researchers helpfully explored a wide variety of scenarios under a variety of both climate change and financial futures, one scenario they did *not* compare was coastal retreat. The goal was to use technology and building codes to make New York City more resilient *where it currently sits*.

This *Science* study is typical in encapsulating a widespread US reaction to climate change: If our environmental circumstances are changing in ways we don't like, we can (and, implicitly, should) simply engineer our way around them. Thomas and Patricia Thornton have labeled this cultural narrative the "Technofix Earth Engineers" narrative, arguing that its proponents "present the Age of Humanity less as a looming crisis than an engineering and enterprise opportunity, replete with calls for planetary management that put scientific and technical personnel at the helm in creating a 'good Anthropocene.'"[38] Or, more bluntly, technology will save us.

The "Technology Will Save Us" climate change narrative is an extension of a "Humans as Controlling Engineers" narrative that, as Chapter 2 will discuss in more detail, was part of a more general post–World War II environmental management narrative that reflected American faith in technology and engineering. This narrative underlies much of our current environmental natural resources law and policy, so we cannot consider the "Technology Will Save Us" cultural narrative about climate change new or surprising. However, it *is* important to distinguish a "Technology Will Save Us" cultural narrative mentality from productive uses of technology in a climate change era. No one doubts that skillful employment of many technologies will be a critical component of both climate change mitigation and climate change adaptation. Humans in general and Americans in particular are technological creatures. On the mitigation side, low-carbon energy generation ranging from nuclear to solar to wind to wave energy all depend on significant technological innovation and improvement. On the adaptation side, a panoply of technologies will be aiding adaptation efforts, from increased use of green infrastructure in cities to more refined dependence on dams, air conditioning, irrigation, and a plethora of other existing technologies.

Employing technologies to help us cope with the Anthropocene and its changes, however, invokes a different mind-set from believing that technology will "save" us. Specifically, a response to climate change rooted in the "Technology Will Save Us" cultural narrative seeks to relatively painlessly maintain our current status quo into the distant future. As such, this narrative tends to elide the need for very real adaptive change and the potentially large dislocations that climate change threatens. Instead, proponents of this narrative tend either to treat the symptoms of climate change rather than its root causes (e.g., sea-level rise rather than greenhouse gas emissions) or to posit that some as-yet-unperfected miracle technology (usually a new form of energy production) will allow us to transition painlessly to a zero-carbon economy.[39]

Like the second narrative ("It Isn't Us"), the "Technology Will Save Us" narrative at least acknowledges that climate change exists. Some versions of it even acknowledge that humans are a critical cause of climate change. Moreover, it clearly empowers humans to respond to climate change. This third narrative would thus appear to be a vast improvement over the first two.

However, the "Technology Will Save Us" narrative has two fundamental problems for the law and policy of the Anthropocene. First, a vague bet on unnamed future technologies is not an adequate basis for the law, policy, and adaptation planning we need to *ensure* that we cope with the Anthropocene: technological optimism is not the same as prudent planning and preparation. Second, the "Technology Will Save Us" narrative posits that humans are predictably in control of nature—i.e., both that nature operates according to highly predictable mechanistic rules and that humans can precisely fine-tune their tinkering with natural systems. For example, Erle Ellis argued in a *New York Times* editorial that "we transform ecosystems to sustain ourselves. This is what we do and have always done."[40]

As we have suggested from the beginning of this chapter and will explore more thoroughly in Chapter 3, that is simply not an accurate description either of natural systems or of human participation in those systems. As such, this third narrative creates problems for both the mitigation and adaptation sides of climate change law and policy: while it appears to be far more empowering than the first two climate change cultural narratives, it nevertheless embodies a form of human arrogance and misunderstanding of complex planetwide systems that sets up human intervention for real and potentially disastrous failure.

On the mitigation side, the extreme form of the "Technology Will Save Us" narrative leads to the promotion of geoengineering technologies to cool the planet—aerosol sprays into the atmosphere, orbiting mirrors to reflect solar radiation, iron fertilization of the oceans to "eat" carbon dioxide.[41] At the very least, geoengineering raises a whole host of risks and a series of political and legal hot potatoes.[42] Moreover, geoengineering technologies are largely unproven technologies, especially at the planetary scale, making geoengineering a planetwide and potentially costly experiment,[43] especially considering that most geoenginecring techniques create risks of unexpected consequences in complex multi-scalar systems. Geoengineering projects thus repeat the human hubris that has attended many much smaller-scale attempts to manipulate nature. This time, however, the fate of the entire planet hangs *intentionally* in the balance.[44]

In addition, geoengineering techniques cannot address some of the critical ecological problems that are the direct result of increasing greenhouse gas concentrations in the atmosphere. The most important of these is ocean acidification—that is, the lowering of the ocean's pH as ocean waters absorb excess carbon dioxide from the atmosphere.[45] According to scientists, even the geoengineering techniques currently being proposed to remove carbon dioxide from the atmosphere aren't enough to save the ocean.[46] As such, the geoengineering manifestation of the "Technology Will Save Us" cultural narrative deflects attention away from some of the very real reasons that we need to reduce the concentrations of carbon dioxide and other greenhouse gases in the atmosphere.[47]

In the adaptation context, the "Technology Will Save Us" narrative can convince us that we can adapt our way through climate change—i.e., that climate change adaptation will be "enough," allowing us to avoid fundamentally changing our lifestyles. To be sure, climate change adaptation is an intensely technological endeavor, and both international and US agencies have been compiling guidebooks of these techniques.[48] Again, however, this cultural narrative is not about just *using* technology; rather, it counsels us that technological adaptation can stave off significant ecological and social-ecological change—for example, that New York City can, with a few building code adjustments, architectural changes, and especially coastal barriers, continue to functional indefinitely (forever?) as it currently does. Notably, the *Science* study for New York City advocated waiting to build the barriers until it was clear that climate change was in fact going to be pretty bad. While acknowledging uncertainty, this approach thus makes light of the Anthropocene's potential for complete unpredictability—the possibility of sudden and catastrophic regime shifts in the planet's climate or other kinds of sudden changes in planetary realities, such as a short-term collapse of the destabilizing Greenland and Antarctic ice sheets, neither of which is behaving precisely as scientists originally expected. The complex planetary systems of which we are a part don't always exhibit linear change or even change within bounded uncertainty.

We need a narrative that acknowledges that our carefully planned technological responses to climate change not only might not be enough but also might not address what turns out to be the fundamental problem for a given SES. Coastal cities again offer a good example. For many coastal cities, it may turn out that saltwater contamination of water supplies (surface water like rivers or groundwater aquifers) becomes the sea-level rise impact that most challenges, or even defeats, the city's adaptation efforts, long before inundation of the city itself becomes a real problem.[49] Nevertheless, coastal climate change adaptation efforts in the United States still emphasize resistance—coastal armoring and sea walls—over dealing with problems like loss of water supply that may necessitate retreat.[50]

These discussions privilege one climate change coastal problem—sea-level rise and coastal inundation[51]—at the expense of other less dramatic but often more determinative climate change adaptation issues. At the same time, they also ignore the potential for sea-level rise to overwhelm even the most seemingly extravagant of coastal technologies. More generally, by focusing adaptation efforts on human control and minimizing disruption and displacement, the "Technology Will Save Us" narrative can actually obscure significant risks to human health and human life, disempowering more effective adaptation options while simultaneously overpromising the effectiveness of specific technology-based adaptation efforts.

It's the End of the World as We Know It

The apocalyptic narrative is deeply embedded in American culture. Indeed, in *The Last Myth*, Mathew Barrett Gross and Mel Gilles ably traced the history of this narrative from Jewish religious tradition through several centuries of history to the United States, concluding that "in America, everyone believes in the apocalypse. The only question is whether Jesus or global warming will get here first."[52] In the early twenty-first century, "Americans increasingly have turned to apocalyptic metaphors to explain and understand a world and nation that look radically different from just a decade ago."[53] However, as Gross and Gilles also note, "conflating our expectation of the apocalypse with the issues before us is an error: the apocalypse is a belief, while the challenges facing us are real."[54]

A climate change apocalyptic narrative dovetails neatly with existing apocalyptic narratives in US culture, both environmental[55] and not. Some of these existing cultural narratives, for example, are religious,[56] and it is worth noting that some churches have embraced climate change as the path toward the Second Coming, possibly impeding efforts to deal with climate change.[57] However, as Michael Burger and others have already pointed out,[58] the United States also has a strong cultural tradition of secular apocalyptic narratives, including in connection with the environment. Importantly, the current generations of "senior decision makers" in the United States can still remember the Cold War and the always-present threat of nuclear annihilation. A childhood of foreign relations policy based on "mutually assured destruction"[59] and the resulting culture of constant fear make it particularly easy for those of us who grew up between the 1950s and 1980s to frame climate change as another potential apocalypse.[60] Indeed, the apocalyptic narrative so dominates US culture that its translation into climate change was probably inevitable, and that translation has now been expressed in everything from popular movies like

The Day after Tomorrow to predictions of the end of human civilization ranging from the *Bulletin of the Atomic Scientists*' "Doomsday Clock" to twenty-first-century versions of the Christian apocalypse.[61]

With this fourth narrative, US climate change cultural narratives flip from denial to overdeterministic acceptance of the worst-case scenario. Ironically, this cultural narrative is, as a result, as disempowering a response to climate change as the first two. Moreover, the fact that "Technology Will Save Us" and "It's the End of the World as We Know It" climate change narratives coexist in US culture is itself a quintessentially American, if somewhat schizophrenic, response. As Gross and Gilles note:

> Indeed, if America is exceptional in one indisputable way, it is that America is the only nation to be founded on *both* the competing ideals of apocalypse and progress. This bifurcated founding is what makes us at once idealistic—known the world over for our friendliness and optimism—and fatalistic—possessed more strongly than many nations by the sense that doom lies just around the corner.[62]

Between this idealism and fatalism, however, we need to find a cultural narrative of realism that can explain that we are in some but not total control of our relationship to nature and environmental processes.

The fourth climate change narrative has a particularly fatalistic variation to it, what might be called the climate change carpe diem narrative—or, to again emphasize the pervasiveness of US apocalypticism, the "Party Like It's 1999" narrative. Examples of this narrative variation are not yet as extensive as they probably will become, but one of the most prominent to date came in response to scientific research published in May 2014 that the collapse of the West Antarctic ice sheet was "inevitable."[63] Scientists originally hedged that full collapse could take several centuries, although studies published since then have almost uniformly documented that Antarctic ice is melting and collapsing much faster than expected, requiring upward adjustments in expected sea-level rise both by 2100 and over the next couple of centuries.

This is information that should prompt worldwide adjustment in coastal management and planning. Instead, *Forbes* magazine chose to feature the conclusion of a group of economists, summarized in the heading "If Antarctic Melting Has Passed the Point of No Return, We Should Do Less about Climate Change, Not More."[64] There are many things that are objectionable about the economists' conclusion, but the aspect that is most dangerous is the assumption that once *some* changes become inevitable, *all* change is inevitable. The apocalyptic narrative, again, overdeterministically insists on fatalistic extremes, whereas the Anthropocene itself resonates along a spectrum of uncertainty.

The most positive formulation of the "It's the End of the World as We Know It" narrative is what Thomas and Patricia Thornton have labeled the environmental jeremiad of the Anthropocene, a moral admonition "that planetary limits are being irresponsibly transgressed by human activity, the footprint of which must be reduced in order to live sustainably within planetary boundaries."[65] This "call to reform" version of the fourth American climate change narrative appears often in environmental news media. For example, two months after *Forbes*'s carpe diem response to the collapse of Antarctic ice sheets, *Forbes* contributor Eric Mack seized upon the potentially long time frame of that collapse to argue that "it's time to finally take the need to reduce climate change emissions seriously while also developing realistic plans for adapting to a warmer, wetter planet. This week's news could mark the end of the world as we know it today, but that should be seen as an opportunity to build a better one."[66]

However, the jeremiad variation is still rooted in fear of destruction, not in human empowerment, limiting its usefulness as a cultural narrative for climate change. As multiple social scientists have emphasized, "'fear framing' or risk-focused appeals to motivate public support of climate change policies have proved largely ineffective at triggering behavioral shifts. . . . 'An excessive focus on negative impacts (i.e., a severe "diagnosis") without effective emphasis on solutions (a feasible "treatment") typically results in turning audiences off rather than engaging them more actively.'"[67]

The truth is, really bad things have happened to human societies in the past, such as the plague that ravaged—and fundamentally changed, it should be noted—feudal Europe. As Gross and Gilles point out, however, scale matters in your perception of these history-changing events. While they are clearly destructive and painful at the scale of individuals, social classes, and even individual nations, from a broader perspective these events "don't end the world—they renew it."[68] So, while the inside of a transformation is not necessarily the most pleasant place to be, we need a cultural narrative that can recognize and name that moment *as* a transformation occurring right now, not as the ending of everything we hold dear.

A New Narrative

We need a new narrative that tells us how to live in the Anthropocene. How humans understand and frame the Anthropocene—this new world of continual, unprecedented, multiple-sector, multiple-scale, and often unpredictable change—matters both to how we experience that change and to how well we interact within SESs. These interactions include how we shape and reflect our experiences through law and policy. The climate change narratives that have emerged so far, however, do not create effective frameworks for either social or legal adaptation, in part because they are so disempowering.

Of course, humans also cannot react to the Anthropocene with a tabula rasa; our future interactions with ecological change will be mediated by the existing structures of SESs. As the US Global Change Research Program reported in 2014, "Climate changes interact with other environmental and societal factors in ways that can either moderate or intensify these impacts."[69] Nevertheless, we believe that the cultural, scientific, and ultimately legal narratives that Americans choose to adopt can still make a considerable difference in how we experience the Anthropocene, especially in how we reenvision the role of environmental and natural resources law and policy in mediating our changing existence within natural systems.

Climate change and the Anthropocene are necessarily large and complex topics, and this book necessarily must limit the subjects it chooses to discuss. This book is about empowering US climate change adaptation projects by offering a new cultural narrative—a new framework for coping with the Anthropocene. It embraces the "radical middle" course of acknowledging that humans are not *fully* in control of natural systems and SESs but nevertheless still possess significant ability to influence those systems and the future directions they take. Before discussing these new narratives, however, we turn in Chapter 2 to an exploration of how we got where we are—the historical narratives of Americans' relationship to their environments that undergird both our current environmental and natural resources laws and our responses to climate change.

CHAPTER TWO

Narrating Our Relationship with Nature

In Chapter 1, we examined how we narrate the phenomenon of climate change as one illustration of why new cultural narratives for the Anthropocene are necessary. This chapter steps back to examine how American culture—and especially our environmental and natural resources law and policy—more generally narrate humanity's relationship to nature, particularly in terms of sustainability and sustainable development. Specifically, it examines three main cultural narratives incorporated into US environmental and natural resources law to date: Manifest Destiny, tragedy, and sustainability. Unfortunately, like the climate change narratives, none of these cultural narratives give us an effective psychological, social, or legal framework for dealing with the changing world of the Anthropocene. As a result, we need both a new cultural narrative and a new framework for environmental and natural resources law. In Chapter 3, we suggest what those might be.

Cultural narratives and the law interact in complex relationships. As Christine Lorillard has observed, "It has become an axiom . . . that law and culture intersect and influence each other. It has also become almost axiomatic that what the law attempts to dictate, culture may not allow to happen."[1] Thus, while changes in the law can promote changes in the surrounding culture, such as in court-issued racial desegregation orders, law cannot get too far ahead of the broader culture without resistance and violence—as was also true of desegregation. More commonly, law reflects changes that have already occurred in society, such as when courts and legislatures incorporated the more general increased cultural attention to consumer protection into safety standards and landlord-tenant law. Either way, cultural narratives are critical to the on-the-ground success of new laws and policies.

Legal theorists often emphasize how important narrative is to law. At a basic level, practitioners exhort the importance of narrative and story-telling in legal persuasion, especially in court and to juries.[2] At a deeper level, Randy Gordon argues that "narratives often stand in the formative background of laws."[3] Thus, cultural narratives can directly influence the exact scope and content of law.[4] In the context of environmental and natural resources law in particular, Professor Michael Burger documents that "stories have played a particularly important role in the development of environmental and natural resources law in the United States."[5] Moreover, existing law can shape how we narrate a changing environment. Laura King argues, for example, that when plaintiffs began figuring out how to sue big oil companies and other industries to address climate change, they created "a new legal narrative [that] soothed psychological chaos and initiated problem-solving by giving shape and in particular by assigning agency to an amorphous problem."[6] In other words, fitting climate change into familiar legal narratives—oil companies are causing a nuisance by promoting greenhouse gas emissions—provided concerned citizens with a sense of empowerment and control in dealing with the apparent chaos of climate change.

For a variety of reasons, however, these old narratives and causes of action are unlikely to provide the US legal system with a comprehensive framework for climate change.[7] One major problem, for example, is that it is difficult to blame just a few actors for causing climate change, no matter how big the corporations might be. Thus, while the *sense* of empowerment conferred through existing legal narratives may be great, *actual* empowerment to deal with climate change and the Anthropocene is far more limited. Many of these lawsuits have not been successful. From this perspective, we argue that these climate change lawsuits constitute just one example of how the United States is falling victim to a widening cultural narrative gap that is becoming a critical barrier to effective natural resources law and policy for the Anthropocene.

This chapter describes the procession of environmental management paradigms that have dominated narratives of humans' relationship to nature in the United States. As noted, these divide into three different types of cultural narratives: Manifest Destiny, tragedy, and sustainability. These three narratives have characterized much of the nation's thinking about environmental change, and, as a result, each is reflected in our current laws and policies. While there is substantial overlap among these narratives in terms of their influence over time, we present them historically, arguing that each narrative is currently informing environmental discourse in various ways.

The Manifest Destiny Narrative

The Manifest Destiny narrative derives from a period in US history beginning in the 1840s. Although it is often referred to as the "frontier era," Jack Wright more accurately describes this period as the "disposal era" of Euro-American expansionism.[8] Homestead laws and other legal mechanisms designed to facilitate rapid settler colonialism characterized this era, when about 20 percent of the public domain was converted into private ownership.[9] As Professor Wright explains, these efforts were predicated upon two myths facilitating the concept of Manifest Destiny—that the land was unoccupied and that it was ecologically pristine.[10] These underlying assumptions justified the dispossession of native people and the conversion of land into capital.

Manifest Destiny itself expressed the belief that it was Anglo-Saxon Americans' providential mission to expand their civilization and institutions across the breadth of North America. While its exact origins are unclear, scholars often point to the words of John Louis O'Sullivan, an American columnist and editor, who used the term in 1845 to promote the annexation of Texas and the Oregon Country to the United States, stating that it was "our manifest destiny to overspread the continent allotted by Providence for the free development of our yearly multiplying millions."[11] From this narrative, two important themes emerged: (1) Euro-American territorialism was inevitable, and (2) it was justified in order to expand the influence of democratic ideals and economic freedom.[12] This narrative is clearly depicted in Figure 2.1, the iconic painting *American Progress* by John Gast in 1872. Columbia (a common early personification of the United States before she was displaced by the Lady Liberty) leads civilization west. She holds a book that "represents learning and knowledge and brings with her the telegraph, the railroad and countless settler colonists."[13]

Manifest Destiny's complex cultural combination of dispossession, democracy, and capitalism remains an important part of the American imagination today. In his work summarizing the Dust Bowl era in the United

Figure 2.1. American Progress, *by John Gast. The United States, as personified by Columbia, is depicted leading western expansion in a mission to expand its civilization and realize its Manifest Destiny. Columbia represents an expansion of a civilized society that brings American settlers as well as "modern" technology and knowledge to cultivate the West.*

States, environmental historian Donald Worster provides insight into this cultural narrative.[14] The Dust Bowl was arguably the largest human-induced ecological disaster in the history of the United States. Aggressive farming practices following the homestead era cultivated millions of acres of prairieland in the Midwest. In less than fifty years, a combination of severe drought coupled with decades of extensive farming without any techniques to avoid erosion created a period of severe dust storms, causing major ecological and agricultural damage throughout the 1930s. Worster explains that the Dust Bowl was the direct result of the underlying cultural narrative of the homestead era, which he describes as based on three main assumptions: (1) nature is capital; (2) humans have a right to use this capital for self-advancement; and (3) social order should permit and encourage this continual increase of personal wealth.[15]

This formula facilitated Euro-American expansionism during the nation's early history, and it is still with us today. It is characterized by notions of dominion and stewardship informed by deeply embedded Judeo-Christian beliefs about the origin of life and the role of humans within the larger religious scheme.[16] While the frontier era was particularly indicative of this mind-set, the narrative of Manifest Density remains an animating force in much natural resource management. For example, concepts like ecosystem services and natural capital are often employed today as metrics for environmental value. They are imbued, however, with what is perhaps Manifest Destiny's deepest cultural assumption of all: nature is here for us.

Here, we invoke the concept of Manifest Destiny in a broader and more ironic sense. The narrative that humans (especially Europeans) are destined to dominate and control the Earth's resources is part of the underlying belief system that created the need for the modern environmental era. Manifest Destiny dovetails directly into the view that human beings are the controlling engineers of nature, an environmental narrative discussed in more detail below. However, in acting out that narrative of domination and control over the course of two centuries, humans created the Anthropocene, revealing the deep tensions that had in fact always existed between humans' ability to significantly influence natural systems and humanity's ultimate inability to fully control those systems.

The nineteenth century's railway and telegraph were only the beginning in terms of emerging technologies that facilitated the human capacity to alter the earth system at an unprecedented pace and scale. In its 2005 Millennium Ecosystem Assessment, the United Nations (UN) concluded: "Over the past 50 years, humans have changed ecosystems more rapidly and extensively than in any comparable period of time in human history, largely to meet rapidly growing demands for food, fresh water, timber, fiber, and fuel."[17] It further concluded that while these "changes that have been made to ecosystems have contributed to substantial net gains in human well-being and economic development, . . . these gains have been achieved at growing costs in the form of the degradation of many ecosystem services, increased risks of nonlinear changes, and the exacerbation of poverty for some groups of people."[18]

The Manifest Destiny narrative encouraged European settlers in the United States to dominate the new nation's natural resources, from land to water to wildlife and fish. Once these settlers had successfully altered the landscape and its resources, however, the emerging negative consequences helped a new environmental narrative to emerge: the tragedy narrative.

The Tragedy Narrative

The tragedy narrative begins in the United States after World War II. The dramatic end of the war and the use of nuclear weapons on the Japanese communities of Hiroshima and Nagasaki were turning points in human history for many reasons, among them the proof that humans could make and use weapons of mass destruction. Thus, the end of war resonates in two valences through American culture. On the one hand, there was a sense of achievement. Our new technological capacity to change our world was viewed with optimism and a sense of opportunity, carried forward most obviously in the increasing exploration of space and the 1969 moon landing. On the other hand, this new capacity filled us with fear. Two entire generations of US children would grow up with the Cold War and recurring exposure to narratives of a nuclear apocalypse, from the movies *Dr. Strangelove* (1964) to *War Games* (1983) and a host of books, movies, and actual political crises in between.

World Wars I and II also developed something arguably more culturally influential than nuclear weapons—they brought us better living through chemistry. The proliferation and use of chemical compounds that occurred both during and following these wars changed our world. Unlike nuclear weapons, chemical improvements at first seemed to have little or no downside. After World War II, however, the dark side of humans' technological interactions with nature and the possible fragility of our environment became increasingly obvious.

While many developments during this period are important, historians generally identify three events as the iconic precipitators of both the tragedy narrative and the American environmental movement, all three of which Ted Nordhaus and Michael Shellenberger emphasize in their book *Break Through: From the Death of Environmentalism to the Politics of Possibility*. The first is the proliferation of pesticides and herbicides in American agriculture and Rachel Carson's subsequent book *Silent Spring*.[19] *Silent Spring* showed us the darker side of better living through chemistry and raised the alarm regarding the unintended consequences of pesticides. The second event was the worldwide distribution of the first view of Planet Earth itself—a photo called "Earthrise" taken by the *Apollo 8* crew in December 1968 (see Figure 2.2).[20] "Earthrise" became an iconic image of the delicate fragility of our planet in the utter blackness of space. The last precipitating event took place a few months later, when in June 1969 an oil slick and debris in the Cuyahoga River in Cleveland, Ohio, caught fire.

Figure 2.2. "Earthrise." Taken by astronaut William Anders during the Apollo 8 *mission, in 1968. "Earthrise" was the first color photograph of Planet Earth. The photograph gave people a new perspective of Earth and became one of the most influential environmental photographs, helping to spawn the environmental movement.*

While this fire was neither the first nor the worst of fires that had occurred in polluted US rivers, it was timed perfectly to draw national attention to water quality and other environmental pollution problems.

All of these events—Carson's book, the "Earthrise" photo, and the Cuyahoga River fire—came at a time when Americans felt deeply vulnerable but also incredibly optimistic. It was this combination of concern and idealism that gave birth to the environmental movement. Americans could no longer avoid the fact that their willingness to engineer ecosystems in the United States came with environmental consequences—dust bowls and exhausted soils in farm lands;[21] the loss of salmon runs in the Pacific Northwest[22] and many parts of the Northeast;[23] polluted waters throughout the United States;[24] and increasing numbers of increasingly imperiled species.[25] The tragedy narrative ushered in the "regulatory era" of the

1970s—a suite of environmental laws that took an ambitious, prescriptive, and control-based approach to environmental management.

While there was a growing fear of our newfound capacities to alter our world, faith in the ability of science and technology to make the world a better place also continued. Importantly, neither the Manifest Destiny nor the tragedy narrative is a single-valence story; instead, both resonate across multiple narrative options. As such, the Manifest Destiny and tragedy narratives of environmental management form more of a continuum than a dichotomy. We tell ourselves multiple stories about our relationship to nature, resulting in different kinds of environmental laws and policies. Michael Burger argues that the development of environmental jurisprudence in the United States has been "something less rational" than what is normally argued—"namely, [it has been] an iterative response to recurrent and competing stories that seek to instantiate competing environmental narratives."[26]

Continued optimism about humans' ability to make the world better led lawmakers to imbed and modernize the Manifest Destiny narrative into a "command and control" management approach within the tragedy narrative that continues to assume our ability to engineer our way out of environmental problems. This command and control approach is in large part a continued manifestation of the technological exhilaration that the United States experienced in the wake of World War II. As such, these statutes still incorporate a form of the Manifest Destiny narrative of human dominance of nature, except now applied to protection of human health and the environment rather than to environmental domination per se. Thus, while an emerging tragedy narrative *prompted* the proliferation of environmental laws in the United States, the Manifest Destiny narrative still strongly informs the structure of those laws and the view of nature embodied within them.

According to these laws, humans have the ability to manage and transform ecosystems to promote the values that humans choose to prioritize.[27] In essence, if humans broke it, humans can fix it. Or, from perhaps a more nuanced perspective, if human priorities for particular ecosystems had changed from pure domination and exploitation to protection and concern for human and ecological health, there was nothing in nature itself to prevent humans from reengineering the relevant ecological systems to suit these new priorities.

As evidence of this assertion, a number of federal pollution control statutes create regulatory regimes that are grounded in human technologies and "technology-based" regulatory standards—the Clean Water Act's effluent limitations,[28] the Clean Air Act's emissions standards,[29] and the Safe Drinking Water Act's maximum contaminant levels.[30] The pervasiveness of this narrative is also evident in the number of federal environmental and natural resources statutes that pursue preservation and restoration as prominent goals, implicitly and explicitly assuming the ability of human managers to return ecological systems to and then keep them within human-defined desirable states of being.[31] For example, the Clean Water Act's overall purpose incorporates both goals, seeking to "restore and maintain the chemical, physical, and biological integrity of the Nation's waters."[32] Both the Comprehensive Environmental Response, Compensation, and Liability Act (CERCLA or Superfund)[33] and the Oil Pollution Act[34] allow governments and Tribes to collect natural resources damages for ecosystems impaired by releases of hazardous substances and oil spills, respectively, and the basic measurement of those damages is the cost of restoring the area to pre-spill or pre-release conditions—a fairly explicit incorporation of the "if humans broke it, humans can fix it" mentality. Similarly, treatment, storage, and disposal facilities regulated under the Resource Conservation and Recovery Act must undertake corrective actions if their activities contaminate land or groundwater, restoring those sites to pre-contamination status,[35] while the Surface Mining Control and Reclamation Act seeks to ensure that mining operations restore the disturbed landscape to something approaching its pre-mining condition.[36] The overall goals of the Endangered Species Act are to prevent the extinction of imperiled species and to restore them to populations that ensure that each species will thrive, often relying on habitat modifications to achieve those goals.[37] However, the Endangered Species Committee, or "God Squad," has express legal authority to allow a species to go extinct if "higher" human needs so demand.[38] Together, these provisions clearly project that the exact fate of imperiled species is a human engineering decision.

Unfortunately, despite the pervasiveness of the "command and control" approach, environmental and natural resources laws are ill-equipped to take on the next generation of environmental challenges. Attempting to optimize and control natural systems to suit human priorities "does not work as a best-practice model because this is not how the world works."[39] Contrary to current ecological science, this approach

> assume[s] that ecological change is predictable and that human impacts are generally reversible. Predictability is what makes human use of natural resources manageable and ecological preservation possible. If regulators can predict how a species, resource, or ecosystem will respond to changes in human impacts (more or less pollution, more or fewer people, more or fewer vehicles, more or less habitat destruction), they can manage that species, resource, or ecosystem to the human-determined functionality or productivity goal. Thus, we require drinking water contamination to be below maximum contaminant levels, manage fisheries for maximum sustainable yield, regulate air pollution to eliminate human health risks, and manage public lands to achieve sustained yield of several products and services. Reversibility, in contrast, presumes that undesirable ecological change can be undone. While some of the exceptions to this assumption are obvious—extinction of species, for example—the whole concept of environmental restoration depends upon it.[40]

Both ecologists and the Intergovernmental Panel on Climate Change (IPCC) have made it clear that predictability and reversibility will become increasingly unlikely in our climate change century.[41] Even without climate change, natural systems continually change in complicated ways, generating complex feedback loops across scales and among systems that lead to unpredictable results. As Daniel Botkin argued in 1992 in *Discordant Harmonies,* there is no such thing as the "Balance of Nature."[42]

Climate change, of course, has become an additional reason why this tragedy narrative of human control no longer works. We have fundamentally shifted the function of the planet to serve human priorities. In so doing, however, we have set in motion any number of positive feedback mechanisms that are accelerating the changes that we and the ecosystems that we depend upon are experiencing. As a result, whatever we *thought* we understood about ecosystems' responses to human technological interventions, climate change fundamentally challenges Americans' ability to effectively narrate, and hence effectively influence, our evolving relationship to evolving natural systems.

It is important to remember, however, that climate change exacerbates rather than creates the disjunction between environmental and natural resources laws and ecological function. Humans have never been able to assert complete control over ecosystems and expect desirable results indefinitely, because we just don't know enough about those ecosystems and their ever-changing, multi-scalar complexity. Climate change does, however, make this disjunction far more visible while simultaneously demanding a change in that framework.

Gone are the days when a scientist like Rachel Carson could carefully review the literature on an environmental concern, isolate the cause, and raise the alarm in a way that results in meaningful change. The global climate change reports from the IPCC over the last two decades provide one of the most saddening examples of this shift.[43] Despite a scientific consensus that anthropocentric causes are the critical driver of climate change, policy efforts have continued to languish.[44]

Carson's formula of "science + fear = change"—a recipe that is still the main strategy embraced by the American environmental movement today—has realistically not worked for decades. The reasons for this failure are many, but there are three worth highlighting. One is that US citizens no longer have the sense of optimism and prosperity that they did after World War II.[45] As a result, we are collectively less willing to take on environmental problems, fearing that doing so would harm economic growth. Another reason is that we no longer have the same relationship with and trust in the federal government, which historically has been the primary source of the "change" aspect of Carson's formula. Instead, increasingly, the federal government (and "government" in general) is viewed with skepticism and suspicion.[46] Finally, this formula is no longer effective because a fear-based discourse tends to have a limited shelf life and a narrow window of opportunity. Long-term environmental and natural resource issues like global biodiversity loss and climate change do not lend themselves to this type of rhetoric.

On a deeper level, the tragedy narrative is problematic because it reflects an underlying belief that humans are *separate from* nature. In this way, as noted, it is still tied to an underlying assumption of the Manifest Destiny narrative. However, while nature is still viewed as subject to our use and control, the tragedy narrative casts humans not as entitled recipients of wealth but as an outside influence that is messing things up. Bill McKibben, one of the most influential environmentalists of our time, provides the quintessential example of this narrative in his book *The End of Nature,* in which the end of nature is human-influenced climate change and other human-based events that make the influence of human beings felt everywhere on the planet: when humans are everywhere, nature is nowhere.[47]

This notion of separateness is a mistake in our perception of reality—an ontological misstep with serious consequences. By placing ourselves outside of nature, we animate both the ideas that nature is "ours" to use (the Manifest Destiny narrative) and what Professor Sarah Krakoff has described as a narrative based on the idea that we are "saving the environment from people and preserving pristine places from contamination."[48] This view pathologizes our current condition, characterizing humans as the main agent of disease/destruction. The challenges of the Anthropocene and its recognition that humans are participants in the environment require new cultural narratives that can more accurately and helpfully describe our complex relationship to natural processes.

The Sustainability Narrative

The third historical narrative informing environmental management is the sustainability narrative. This narrative focuses less on problems and fears and more on finding a more balanced way to manage the impacts associated with resource consumption and associated environmental woes. Building on the discussion in Chapter 1, in its most general definition, "sustainability" refers to the long-term ability to continue to engage in a particular activity, process, or use of natural resources. To some extent, this narrative gained initial influence in the United States in the 1970s, when "multiple-use-sustained-yield" became the management principle for many natural resource management regimes and early UN conferences began to articulate development goals.[49] It is fair to say, however, that sustainability as an environmental management narrative really gained steam much later, when the international community embraced sustainable development at the 1992 UN Conference on Environment and Development in Rio de Janeiro, incorporating it into both the Rio Declaration and Agenda 21.[50] This conference, known as the "Earth Summit," was the same conference during which world leaders opened for signature the UN Framework Convention on Climate Change. Professors Dernbach and Cheever have outlined the origins of sustainability and sustainable development, noting these concepts' integration into numerous international treaties, including the Framework Convention on Climate Change and the Convention on Biological Diversity.[51]

Like all of the narratives we discuss, the sustainability narrative rests on underlying assumptions about social-ecological systems (SESs)—assumptions that no longer match ecologists' evolving understanding of complex planetary systems, particularly in light of climate change. Specifically, the sustainability narrative tends to assume that humans: (1) know what can be sustained, and how; and (2) have the capacity to maintain complex systems within some envelope of stationarity, balanced equilibrium, or bounded variability. The personal finance and simple timber harvest examples from Chapter 1 show both of these assumptions in operation—and the risks they produce because the focus of the sustainability goal (money, timber) is situated within larger complex systems. Even the most carefully planned personal budget can become unsustainable in the face of high inflation or a global economic crisis, while "sustainable" forestry practices are vulnerable to climate perturbations and invasive pest species.

Given this increasingly obvious mismatch with reality, we argue that the sustainability narrative, like the Manifest Destiny and tragedy narratives, has also reached the end of its usefulness. Notably, the pursuit of sustainable development goals has not resulted in effective mitigation of climate change; instead, greenhouse gas emissions have continued to increase, as have resource consumption patterns in terms of pace and scale.[52] Biodiversity loss is also increasing at exponential rates.

In the summer of 2012, the UN held the "Rio + 20" conference, reflecting on the twenty years that had passed since the Earth Summit.[53] In anticipation, the UN Environment Programme issued its Global Outlook report, which Executive Director Achim Steiner summarized by stating, "if current patterns of production and consumption of natural resources prevail and cannot be reversed and 'decoupled,' then governments will preside over unprecedented levels of damage and degradation."[54] The report emphasized the increasingly likely possibility of large-scale irreversible change, concluding that as human pressures on the Earth system accelerate, critical thresholds at various scales are quickly being approached or, in some cases, have already been exceeded. Particular emphasis was placed on nonlinear change—impending social and ecological thresholds that, once crossed, would prove irreversible.

The report reflects a growing consensus that "stationarity" (the idea that natural systems fluctuate within an unchanging envelope of variability) is dead.[55] Yet any cursory review of environmental programs and associated academic literature reveals that the culture still embraces the sustainability narrative. By definition, sustainability assumes that there are desirable states of being for SESs that humans can maintain. This is a questionable assumption under the best of circumstances, and, in practice, sustainability-based goals proved difficult to achieve even before climate change came on the scene.

Given these facts, it is worth articulating fully what sustainability and sustainable development goals in the Anthropocene actually presume. And so:

In pursuing sustainability in the Anthropocene, humans are asserting that we understand our complexly changing complex world well enough to be able to successfully pursue continuous economic development and social betterment for a human population that is steadily and significantly increasing without compromising the environmental amenities that emerge from healthy natural systems, even though these natural systems are changing at multiple scales in ways that we cannot fully predict.

This assertion is, we submit, an extremely dubious proposition, one grounded in societal hubris rather than ecological reality.

The sustainability narrative offered a noble vision for the future. Moreover, two particularly important elements of the sustainability story are worth moving forward into the Anthropocene: first, that we cannot consider environmental, economic, and social issues in isolation; and second, that *inter-* and *intra*-generational equity must be considered when crafting policy approaches. However, the Anthropocene demands difficult conversations about the trade-offs among economic development, social improvement, and environmental protection, as well as the future costs of present actions.

Rather than driving these difficult policy discussions, the sustainability narrative, at least in the United States, has devolved into a "have it all" discourse grounded in green consumerism.[56] Thus, Michael Burger notes that, while sustainability has been the most influential environmental idea of the last thirty years, its underlying story is a utopian one—one that is at best unrealistic and at worst deceptive:

Sustainability has failed . . . to compel the radical transformation at the core of the countercultural social movement that invented modern environmental politics. Rather than inspire changes in the way we live necessary to actually redress the environmental crisis, the sustainability story brackets big-ticket items like capitalism and consumerism, reifies existing actors and hierarchies, and affirms basic patterns of social organization, production and consumption. In short it is a deceptive story that perpetuates existing power dynamics that are in many respects the causes of global climate change.[57]

Nevertheless, there remains a general reluctance to let go of this narrative. Sustainable development is a more specific incarnation of sustainability. The sustainable development narrative emerged at about the same time as scientists were becoming convinced that climate change was occurring and that humans had something to do with it. While the International Union for the Conservation of Nature (IUCN) dates the concept of sustainable development to its 1969 mandate and the 1972 United Nations Conference on the Human Environment (Stockholm),[58] the 1987 report of the World Commission on Environment and Development (also known as the Brundtland Commission), "Our Common Future," is generally credited with launching sustainable development as an international governance goal.[59] Indeed, that report provided the most commonly used definition of sustainable development: "Development that meets the needs of the present without compromising the ability of future generations to meet their own needs."[60] Sustainable development goals were further operationalized in 1992 at the "Earth Summit," particularly in Agenda 21,[61] and in 2000 sustainable development itself became one of the UN's eight Millennium Development Goals.[62]

Sustainable development can describe either a decision-making process or a substantive goal. We focus mainly on sustainable development as a substantive goal, but we also note that many limitations of the substantive sustainable development narrative also apply to sustainable development as a decision-making framework, particularly in terms of an inability to contemplate or effectively deal with hard ecological limits on human development.

Sustainable development has been defined and redefined various ways.[63] Literally, as sustainable development textbook author Jennifer Elliott has noted, "sustainable development refers to maintaining development over time."[64] However, not all versions of sustainable development tell exactly the same story. Most importantly, pursuers of sustainable development make different assumptions about the relationship between human beings and the environment.[65] These differing assumptions are evident in

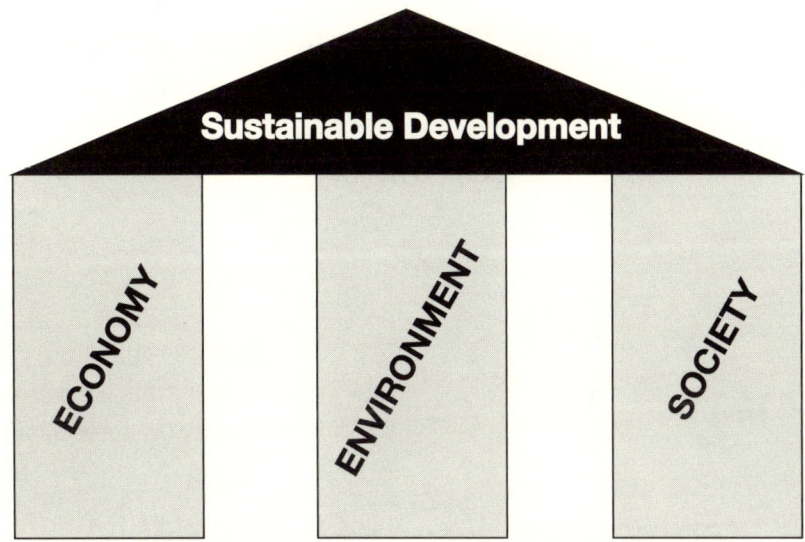

Figure 2.3. The Three Pillars Model of Sustainable Development. The three pillars—economy, environment, and society—are depicted here as separate but equal foundations of sustainable development. This model ignores the connections and trade-offs among the pillars. Moreover, the environment almost always is the center pillar, subtly suggesting that it can be eliminated and sustainable development can still be held up by the economy and society.

three common visual portrayals of sustainable development—the "three pillars" model, the "interlocking circles" model, and the "nested spheres" model—which effectively create and portray three different cultural narratives about the relationships among society and social welfare, the environment, and economic development.[66]

The three pillars model (Figure 2.3) "confirm[s] the need to consider the social, ecological and economic arenas together and equally" to achieve sustainable development, but it does not clearly depict the interconnections among these pillars. In particular, the three pillars model does not acknowledge that both economic development and social well-being, as well as sustainable development overall, each depend upon well-functioning ecosystems that can continue to deliver goods and services. In other words, the three pillars model perpetuates a narrative that social and economic systems can exist and function independently of the environment. Indeed, most portrayals of this model are two-dimensional, with the environment pillar in the middle, subtly suggesting that the sustainable development goal will remain supported even if the environmental pillar evaporates.

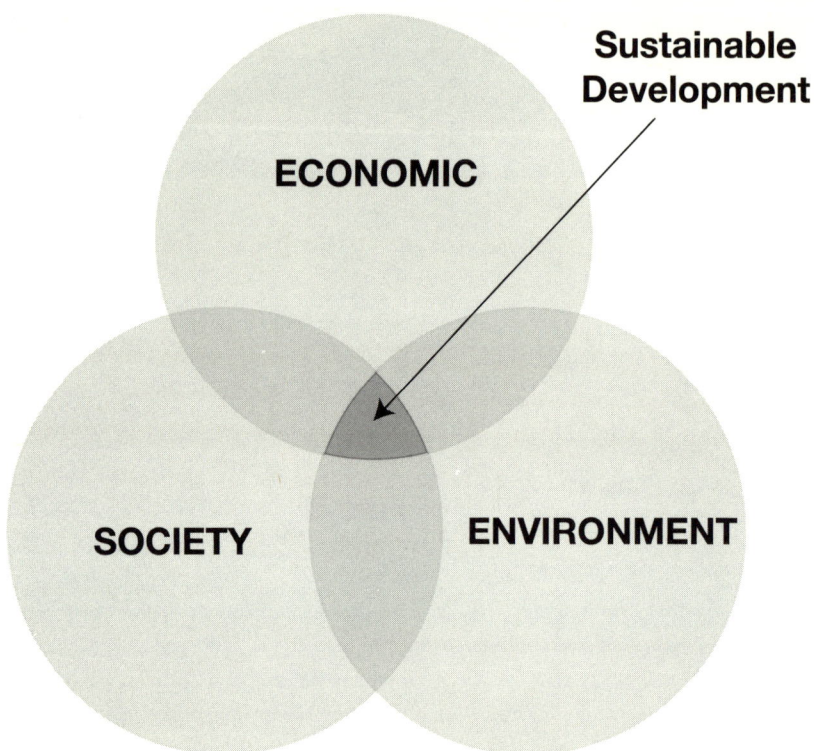

Figure 2.4. The Interlocking Circles Model of Sustainable Development. This portrayal of sustainable development acknowledges that the three components—environment, economy, and society—interact and that there might be trade-offs required in order to achieve the center "sweet spot." However, no place in this model is a bad place to be.

The three pillars model thus embodies a narrative that undermines—albeit subtly and probably unintentionally—the very real and very important fact that all social and economic systems depend on the environment.

Unlike the three pillars model, the interlocking circles model of sustainable development (Figure 2.4) more clearly narrates "the need to integrate thinking and action in sustainable development across traditional disciplinary boundaries and established policy-making departments."[67] Indeed, in its Programme for 2005–2008, the IUCN adopted "the interlocking circles model to demonstrate that the three objectives need to be better integrated, with action to redress the balance between dimensions of sustainability."[68] The middle area of overlap represents the area in which the goals of all three spheres are all maximized—i.e., "the possibility of mutually supportive ('win-win-win') gains" in all three areas (economic, social, and environmental) simultaneously.[69] Moreover, "the small area of overlap relative to the whole sphere portrays the unsustainable nature of much activity, but also opens the idea of the potential to expand this area of positive overlap."[70]

An important but often overlooked aspect of the interlocking circles model is the concept of trade-offs. Specifically, the "sweet spot" in the center representing true sustainable development is the product of many trade-offs among the three areas, suggesting that a balancing rather than "have it all" mentality is absolutely necessary. The interlocking circles model thus also supports the idea that sustainable development requires systems thinking—that is, an approach that acknowledges that social improvement, economic development, and the environment exist as complex interactions rather than isolated arenas. The IUCN, for example, has identified "two fundamental issues" for sustainable development: "the problem of environmental degradation that so commonly accompanies economic growth, and yet the need for such growth to alleviate poverty."[71]

Like the three pillars model, the interlocking circles conceptualizes the three areas of focus (society, economics, and environment) as equal and somewhat commensurable—i.e., it suggests no limits on humans' ability to trade improvements in one sphere (say, economic development) for degradations in another (say, the environment). As such, the interlocking circles model, like the three pillars model, can (again, albeit subtly) support a weak sustainability approach that undermines the environmental "bottom line" of human existence.

Environmental limits do emerge, however, from the third model of sustainable development, the nested spheres model (Figure 2.5). In this model, "the spheres of economy and society are shown as embedded in a wider circle of ecology," conveying "an understanding of environmental limits setting boundaries within which a sustainable society and economy must be sought."[72] This model acknowledges that "activities that damage the functioning of natural systems ultimately weaken the basis of human existence itself."[73] The nested spheres narrative underscores the fact that the societal and economic goals of sustainable development ultimately depend on a rich and well-functioning environment that supplies, at the very least, critical natural capital and preferably more extensive ecosystem goods and services.

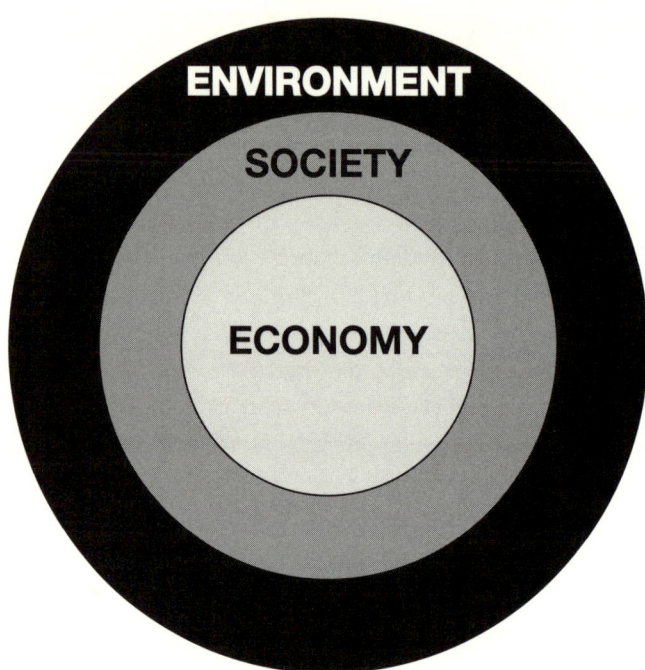

Figure 2.5. The Nested Spheres Model of Sustainable Development. The economy and society are nestled within the constraints of the environment, depicting the importance of natural systems in achieving economic and social goals. The best outcomes for the economy and society result from maintaining a well-functioning environment that can support such activity.

Unfortunately, these nested spheres are difficult to find in the United States. Instead, the existing US narratives of sustainable development do not seriously contemplate the possibility that there might be limits to development, let alone explain either how to pursue sustainable development goals in the face of such limits or how to prioritize goals when ecological limits make the simultaneous achievement of all desired outcomes (continued development, poverty alleviation, increased prosperity, and environmental protection and restoration) impossible. However, this is the reality that the Anthropocene is creating, calling into question the continued viability of all major US sustainable development narratives because they elide unavoidable trade-offs and fail to contemplate that human civilization may be facing the unavoidable decline of its development to date.

Outside of the United States, concerns about the future of sustainable development in the Anthropocene are mounting in light of the increasing challenges that climate change poses. In 2006, the IUCN acknowledged that "the evidence is that the global human enterprise is rapidly becoming less sustainable and not more."[74] As noted in Chapter 1, the IPCC in 2014 concluded that climate change could undermine sustainable development goals. Indeed, the last highlighted message that its Summary for Policymakers conveys is that "climate change is a threat to sustainable development."[75] More specifically, the IPCC now concludes that "climate change poses a moderate threat to current sustainable development and a severe threat to future sustainable development."[76] Although the IPCC still hews to sustainable development as a global goal, it acknowledges that climate change could vitiate that goal. As it notes in its classically reserved tone, "added to other stresses such as poverty, inequality, or diseases, the effects of climate change will make sustainable development objectives such as food and livelihood security, poverty reduction, health, and access to clean water more difficult to achieve for many locations, systems, and affected populations."[77]

Even sustainable development advocates increasingly acknowledge that climate change may undo the whole sustainable development enterprise. For example, Jeffrey D. Sachs, in his 2015 book, *The Age of Sustainable Development,* acknowledges the challenge that climate change poses to sustainable development, emphasizing that "all of our civilization—the location of our cities, the crops that we grow, and the technologies that run our industry—is based on a climate pattern that will soon disappear from the planet."[78]

As the global community begins to acknowledge that climate change poses an increasingly serious threat to sustainable development, other evaluations of humans' use of our environment underscore that those uses are currently nowhere near sustainable. Most of these efforts, like the nested spheres model of sustainable development, acknowledge that the environment eventually limits certain kinds of human development, especially the materialistic and wasteful development that has characterized the last three centuries in the West and that increasingly accurately describes development in other parts of the globe. As the IUCN recognized in 2006, "the uncomfortable bottom line of sustainability is the insight that the biosphere is limited."[79]

Various measures suggest that our consumptive behaviors are surpassing those ecological and planetary limits. For example, environmental "footprint" measures determine the impact of human consumption patterns in terms of the amount of resources needed to meet various human uses. Numerous apps and websites offer individuals opportunities to calculate their carbon footprint, their energy footprint, their water footprint, and so forth. The Global Footprint Network, however, has pursued this kind of consumption measure on a global scale, resulting in a concept of "ecological overshoot." Each year, the Network calculates Earth Overshoot Day, i.e., the date each year on which humanity has used all of the renewable natural resources that the planet can produce in one year. In an exactly sustainable world, humanity would use up the year's resources on December 31 at midnight. Any date prior to December 31 means that humanity has overshot Earth's renewable resource capacity—i.e., that we are living unsustainably. In 2016, Earth Overshoot Day occurred on August 8, meaning that humanity needs more than 1.6 Earths to support its demands on planetary ecosystems.[80] Studies such as these suggest that the human footprint on the world, in terms of consumption, degradation, and waste, is large and extremely damaging, raising concerns that we may be exceeding the planet's capacity to adapt.[81]

Another approach to assessing the ecological limits of development is the planetary boundaries concept. Researchers working out of the Stockholm Resilience Centre have attempted to quantify nine key planetary system functions and then to assess where humanity is in terms of the risk of (perhaps irrevocably) crossing those boundaries. They are concerned that, "for the first time in human history, we may have pushed the planet too far."[82] As Johan Rockström and Mattias Klum, two key researchers for the Planetary Boundaries project, explain, "the groundwork for the planetary boundaries concept rests on more than 30 years of empirical research showing that ecosystems, from local lakes to forest biomes and large ice sheets, can abruptly cross tipping points and irreversibly shift from one stable state to another."[83] Their goal was to identify those systems and tipping points that could put the entire planetary system at risk, in the process defining what they describe as "the safe operating space for humanity on a stable planet."[84] The international team of researchers originally proposed nine planetary boundaries in 2009 in peer-reviewed research, then revised and updated the framework in an article published in 2014 in *Science* and depicted in Figure 2.6. The nine planetary boundaries they identified are: climate change; biodiversity; novel entities, such as invasive species; stratospheric ozone depletion; ocean acidification; biochemical flows (disruption of phosphorus

Figure 2.6. Planetary Boundaries. An Earth system framework pioneered by Johan Rockström from the Stockholm Resilience Centre and Will Steffen from the Australian National University in which nine key planetary system functions are assessed in terms of human activity. Areas toward the center represent human activities that are operating in a safe margin, the lighter bands starting at the second ring out from the center depict human activities that may or may not have exceeded safe margins, and the outer two rings depict where human activities have exceeded safe operating margins. The gray areas with question marks represent areas where the safe margins for human activities have not yet been determined. This diagram thus indicates that biochemical flows and genetic diversity loss currently pose the greatest risks to the planet, although climate change is close behind. Image courtesy of user Ninjatacoshell via Wikimedia Commons.

and nitrogen cycles); land system change; freshwater use; and atmospheric aerosol loading.[85] Of these nine, the researchers consider two to be "core boundaries," because "the climate system and the richness of biodiversity on Earth have a decisive role, on their own, in determining the outcome of the planetary state."[86] Unfortunately, humanity is pushing both of these core boundaries, plus two others, toward their tipping points: The research team concluded in 2014 that humanity has crossed into high-risk zones for biodiversity and biochemical flows and into a zone of uncertainty for climate change and land use change.[87]

It is important to note that these calculations are moving targets and to avoid a Malthusian mind-set. Thomas Malthus famously predicted in 1779 that the earth's population would outstrip agricultural production.[88] He was wrong; innovations in agricultural practices, increased efficiencies, and new technologies changed the predicted trajectory. Since that time, the term "Malthusian" has been used to describe predictions of catastrophe arising because of the human population overshooting resources (for example, Paul Ehrlich's predictions regarding world population growth).[89] Changes in consumption patterns and new technologies can "decouple" economic growth and environmental destruction, which is one of the main reasons why the tragedy narrative has failed.[90] *Relative* decoupling refers to impacts growing at a slower rate than population or total consumption. As one example, water consumption rates per person in the United States have fallen despite an increasing population. *Absolute* decoupling means that impacts are declining in absolute terms.[91] In the United States, reductions in lead air pollution are an example: the absolute concentration of lead in the atmosphere (and in human blood) has dropped precipitously since the US Environmental Protection Agency and Congress outlawed lead as a gasoline additive.

Regardless of specific calculations, research suggests a significant need for environmental and natural resources law and policy to focus far more intensively on decoupling efforts to address climate change mitigation and biodiversity preservation—that is, on incentives and regulatory measures that can sustain human health and a comfortable lifestyle while generating far fewer greenhouse gas emissions and competing less with other species for shared resources. Notably, the planetary boundaries researchers themselves underscore the need for a new cultural narrative for the Anthropocene. Observing that "ever since the industrial revolution, we've had this crazy idea that our actions are without consequences. That we can take nature or leave it,"[92] they conclude instead that "what we need now is a deep rethink, a total mind-shift about the way that our economies should develop within the life-support systems on Earth."[93]

The world needs a new narrative—a positive story about new opportunities for humanity to thrive on our beautiful planet by using ingenuity, core values, and humanism to become wise stewards of nature and the entire planet. The dominant narrative until now has been about infinite material growth on a finite planet, assuming that Earth and nature have an endless capacity to take abuse without punching back. That narrative held up as long as we inhabited a relatively small world on a relatively big planet—one in which Earth kept forgiving all the insults we threw at her. But that is no longer the case. We left that era 25 years ago. Today we inhabit a big world on a small planet—one so saturated with environmental pressures that it has started to submit the first invoices to the world economy: the rising costs of extreme weather events and the volatility of world food and resource costs. We need a new way of thinking about our relationship with nature, and how reconnecting with the planet can open up new avenues to world prosperity.[94]

We need to reevaluate the sustainability narrative and its usefulness for the Anthropocene. Specifically, we must critically examine whether long-term "sustainable" development in the post–Industrial Revolution mold is even still possible and whether we can adequately define "sustainability" as applied to the real world to provide useful guidance in law and policy to natural resources managers.[95] The changes that Earth is experiencing and will continue to experience, we argue, put the knowledge base necessary for true sustainability beyond human grasp.

Climate change interacting synergistically with other anthropogenic stressors to the planet—pollution, habitat destruction, unsustainable natural resource use—means that humans can no longer depend on past environmental conditions to be accurate predictors of future realities—or hope to restore those conditions after they change.[96] Climate change makes obvious that human influences on complex planetary systems are altering—in some cases transforming—SESs. As the IPCC concluded in 2014, "Continued emission of greenhouse gases will cause further warming and long-lasting changes in all components of the climate system, increasing the likelihood of severe, pervasive and irreversible impacts for people and ecosystems."[97] Moreover, "climate change will amplify existing risks and create new risks for natural and human systems."[98] These actual changes and risks of change are also long-term: "Many aspects of climate change and associated impacts will continue for centuries, even if anthropogenic emissions of greenhouse gases are stopped. The risks of abrupt or irreversible changes increase as the magnitude of the warming increases."[99] As a result, ecologists conclude, "increasingly today, and most certainly in the future, the only constant will be change. Surprise is the new normal,"[100] especially because "sudden, unexpected change appears to be the rule rather than the exception in natural systems."[101]

As ecosystems and other planetary systems progress through this new era of continual, complex, unprecedented, and often unpredictable change, "the earth's capacity to yield products for human consumption, to absorb or sequestrate human wastes (especially novel compounds), and to yield ecosystem services are all of them limited. The idea that there is always somewhere to absorb externalities is flawed, and it is a myth of progress that living systems will always recover from human demands."[102] If, as many scientists and the IUCN suspect, we are approaching (or passing) a suite of ecological thresholds and planetary boundaries, humans' options for future development are narrowing, or perhaps disappearing altogether. This potential loss of future options poses risks to SESs that should already be modifying how humans think about development goals. All societies ultimately depend on ecosystems and the goods and services that those ecosystems provide, but climate change directly threatens the current states of most of the world's ecosystems.

The climate change extremes of this new reality, such as the predicted disappearance of island nations as a result of sea-level rise,[103] have been well publicized but not yet incorporated into global development goals. In part, these kinds of extreme, existential threats to island (and Arctic) cultures and nations may not seem generalizable; indeed, they are currently generally portrayed as tragic but somewhat unusual climate change fates for particular kinds of human societies, with the implication that the rest of us will still be able to muddle along in our pursuit of continuous development. Ecological dependence, however, is more insidious than that, and at some point, a society's dependence on a failing or radically changing ecosystem drastically retards, even reverses, economic and social development.

For societies that lose their homelands, food supply, or water supply, this statement does not go nearly far enough. Sustainable development goals—indeed, *any* development goals—presume that the relevant society will continue to have the basic ecological requisites for development—a place to inhabit, a source or sources of food, water that is or can be made potable. Climate change calls those assumptions into question.

The reality of the Anthropocene is that, as planetary systems alter and transform, we will increasingly have only the most tenuous of ideas of how "sustainable" our uses of the planet and its resources might be. The disjunction between the sustainability narrative and our new reality calls for a replacement cultural narrative—as Rockström and Klum suggest, we need a better framework for thinking about our evolving relationship to nature, one that encompasses change and unpredictability as our new normal.

Chapter 3 offers two intertwined threads of a new cultural narrative for the Anthropocene. From ecology comes resilience theory, a more nuanced description of ecological reality that simultaneously provides a better vocabulary and narrative framework for a world of continual change. On the cultural side is the myth of the trickster, that unpredictable agent of change who teaches us both humility and the important lesson that we can, in fact, cope with a changing world.

CHAPTER THREE

Resilience and the Trickster

A New Narrative for the Anthropocene

In Chapter 2, we explored US culture's numerous narratives regarding our relationship to nature and determined that none of them are a good fit for the Anthropocene. At the same time, as we discussed in Chapter 1, the narratives that are emerging about climate change are equally unhelpful, often disempowering.

In our relationship to ecosystems in the Anthropocene, we find ourselves on a middle ground of agency that is characterized by neither impotence nor omnipotence. Unfortunately, as both Chapters 1 and 2 explored, culture narratives in the United States regarding this relationship tend not to cover this middle ground, favoring instead variations on a theme of nearly complete human understanding and control or, conversely, nearly complete negation of human agency. This chapter offers an alternative cultural narrative that *can* help residents of the United States stand more comfortably in the realm of limited agency, uncertainty, and surprise. We base this new narrative on the cultural narrative of the trickster and the scientific theory of ecological resilience.

Resilience theory provides a more complete and nuanced scientific approach to how social and ecological systems function that goes beyond traditional mechanistic and "steady state" models. For that reason alone, resilience theory warrants a reconfiguration of US environmental and natural resources law and policy, which, as already discussed, are grounded in that outdated, mechanistic view of nature. Specifically, environmental management efforts are currently based on an oversimplified view of natural systems in which all impacts are reversible and all damaged ecosystems can be restored. Incorporating resilience theory into law is especially helpful in the Anthropocene because it recognizes "surprise"—the inherently unpredictable nature of social and ecological systems. Resilience theory can therefore help us to better cope with the accelerating change resulting from climate change and its synergistic interactions with other anthropogenic stressors.

In the dominant US culture, however, resilience theory's scientific framework currently lacks a resonating cultural narrative, impeding its assimilation into law and management. To borrow terminology from Robert A. Heinlein's classic science fiction novel *Stranger in a Strange Land*, Americans have a hard time *grokking* (understand deeply, at a gut level) resilience theory because, culturally, Americans don't view themselves as limited agents. The United States is instead a culture of the American dream and radical liberal individualism.

In this chapter, we propose the "trickster" as a means of filling this cultural narrative gap. While generally lacking in Euro-American culture, most other cultures (notably Native American cultures) have a trickster that brings about unexpected change. The trickster[1] playfully disrupts normal life and then reestablishes it on a new basis.[2] Resilience theorists have already acknowledged the connection between resilience theory's narrative of natural systems and the trickster in the concept of panarchy, an element of the theory explained in more detail below. For now, it's worth noting that C. S. "Buzz" Holling and Lance Gunderson coined the term "panarchy" after the Greek trickster god Pan in order to describe the complex, dynamic, and often surprising relationships of social-ecological systems (SESs) across different scales.[3] For resilience to become an effective cultural narrative for the Anthropocene, US culture must embrace this aspect of the theory and acknowledge climate change for the trickster that it is.

Here, we discuss the power of the trickster narrative and frame climate change as one manifestation of the trickster as an agent of change. However, while the trickster narrative is a helpful cultural narrative to adopt in order to cope with climate change, it can only contextualize, rather than operationalize, a new approach to environmental and natural resource law and policy. On the operational end, resilience theory offers the framework for explaining and coping with change in natural systems while simultaneously suggesting a much more productive legal and policy approach for this new era. After explaining the need for the trickster as a cultural narrative, we provide an overview of resilience theory. The chapter concludes by examining the implications this new narrative of resilience for US law and policy.

Climate Change as the Trickster

As Chapter 1 already described in detail, our current narratives do not provide a productive way of thinking about climate change. They are all about either ignoring or resisting change or about giving up. Luckily, a different kind of narrative exists in many cultures that can far more productively frame climate change, allowing for a more productive attitude toward coping with its surprises and transformations: the story of the trickster.

In general, folklore stories like those of the trickster can become powerful cultural narratives for dealing with climate change because they place humans in a different relationship to ecological change[4] than the dominant US narratives do—humans are neither controlling engineers nor victims of natural forces but rather components of a complex system who have a real but bounded ability to deal with its changes. As Thomas and Patricia Thornton have noted, "The tenor and rhetoric of the prevailing discussions of climate change and the Anthropocene are at odds with an alternative heuristics circulating in many indigenous communities that are instead shaped by the shared understanding that humans are but a small part of a relational universe that cannot be fully cognized, much less managed, by any one species."[5]

As Michael Chabon has noted, the trickster "embodies the contingent and in so doing lends it the appearance of necessity."[6] Tricksters are agents of chaos, forces that disrupt normal expectations and sometimes violate important cultural or sacred boundaries.[7] The trickster brings change and crosses thresholds. In Lewis Hyde's words, the trickster "is the spirit of the doorway leading out, and of the crossroad at the edge of town," "the spirit of the road at dusk," and "he is the adept who can move between heaven and earth, and between the living and the dead."[8] "In short, trickster is a boundary-crosser":

> We constantly distinguish—right and wrong, sacred and profane, clean and dirty, male and female, young and old, living and dead—and in every case trickster will cross the line and confuse the distinction. Trickster is the creative idiot, therefore, the wise fool, the gray-haired baby, the cross-dresser, the speaker of sacred profanities. Where someone's sense of honorable behavior has left him unable to act, trickster will appear to suggest an amoral action, something right/wrong that will get

life going again. Trickster is the mythic embodiment of ambiguity and ambivalence, doubleness and duplicity, contradiction and paradox.[9]

It is the trickster's role to test the continuing viability of human categories and to redraw sacred boundaries to suggest new modes of being when prior categories no longer function effectively. His boundary crossing often operates to improve human life, as when tricksters in many cultures steal amenities from the gods for humanity. Examples include Prometheus stealing fire, North Pacific Native Americans' Raven stealing water and daylight, Japan's trickster releasing the arts of agriculture from heaven, and Hermes stealing cattle.[10]

Two aspects of the trickster are particularly important to the Anthropocene. First, the trickster is generally neither good nor evil; he is amoral.[11] As will be true with resilience theory, the trickster is simply a facet of reality, not a moral theory or prescription. It is therefore up to humans to give meaning (or not) to the changes and disruptions that the trickster brings. Second, as humans interact with the trickster and his disruptions, they learn to adapt and change to accommodate the new realities that the trickster brings, helping to ensure their own survival. "Trickster the culture hero is always present; his seemingly asocial actions continue to keep our world lively and give it the flexibility to endure."[12] Thus, when the trickster "lies and steals, it isn't so much to get away with something or get rich as to disturb the established categories of truth and property and, by so doing, open the road to possible new worlds."[13] Indeed, one of the points of trickster tales is that humans and cultures *need* to be able to adapt—"that the origins, liveliness, and durability of cultures require that there be space for figures whose function is to uncover and disrupt the very things that cultures are based on."[14]

While trickster stories exist all over the world and in most cultures, anthropologists tell us that the trickster is almost insistently absent from one prominent culture: the Euro-American culture of the United States.[15] Our best (and fairly pathetic) attempts are the con man and Murphy of Murphy's Law: if anything can go wrong, it will. Notably, Murphy's Law is typically American in framing the unexpected as inherently bad, an unwelcome interruption of planned events that inevitably makes the situation worse. In contrast, many Native American cultures celebrate many trickster tales, focusing on figures that include Coyote, Raven, Spider, and several others. Many of these tales are comic, not tragic, and the changes these tricksters bring are often beneficial or at least harmless to humans. Thus, trickster tales offer a rich, complex, and nuanced set of narratives about how humans can and do experience unexpected disruptions, producing a much more productive cultural narrative for coping with the Anthropocene than those discussed in Chapters 1 and 2.

One aspect of tricksters is to teach humans to use technology, such as fire. As Lewis Hyde documents, tricksters from the Norse Loki to the Greek tricksters to the Native American Raven and Coyote, and many more, devise methods to fish and to trap animals, which humans then adopt.[16] Trickster tales thus acknowledge the role of technology in taking advantage of natural resources. However, trickster stories also acknowledge that technology works both ways: "trickster can also get snared in his own devices."[17] Moreover, trickster tales from around the world document the increasingly complex roles that tricksters can play in human hunting and fishing, evolving out of simple predator-prey relationships (trickster as trapper versus tricksters being trapped) to stories where the trickster (for example, Raven or the Zulu trickster Thlókunyana) learns to steal the bait that humans are using to catch fish or game, creating a third narrative role for these stories.[18] In these stories, there is no final mastery of natural resources. Instead, deployment of technology into nature becomes a continual series of frustrations and adaptations.

At a most basic level, tricksters teach humans to expect the unexpected and that change—good or bad—is just part of life. As such, trickster tales often start with and focus on the most basic of changes (eating and death),[19] but they also often make larger points about humans' interactions with a continually changing reality. For example, in one tale from the Tsimshian of the Pacific Northwest, Raven is hungry and wants a whale that villagers have hunted and brought up on shore. As Raven, he causes a commotion on the beach, then turns himself into a human man to translate the Raven language, telling the villagers that a deadly disease is coming and they have to leave. The villagers do, and so Raven gets the entire whale—and the village—to himself. After he eats his way through the whale, however, Raven creates a slave to move on to another village, playing essentially the same trick on the humans—but also being tricked out of some cod by his new slave. When, at a third village, the created slave tricks Raven out of a sweet crabapple dish, Raven kills him, retrieving the "stolen" food from the slave's stomach. Raven ends that tale by flying away, looking for his next adventure.[20]

The lessons of this tale for the Anthropocene are many. For example, the villagers' completely rational response to a perceived threat of disease turns out to be unnecessary because they have been tricked—but they *are* in fact able to cope with such threats. Raven's greed cheats many others—two entire villages—out of critical natural resources, but Raven ends up alone with his glut of villages that are no longer productive. Raven creates a helper to enhance his ability to access food at new villages, only to have that slave reduce the food available to Raven. Raven's greed for natural resources (food), therefore, condemns him to perpetual solitude and movement while simultaneously destroying the prosperity of many others.

The Miwok, who occupied the foothills of the Sierras in California, and the Crow, who historically lived in the Yellowstone River valley, tell trickster tales more resonant with climate change. In the Miwok tale, Coyote lived in the Village of Darkness but adventurously traveled into the Village of Light, where he saw the chief of that village alternately set the sun and the moon on their travels, bringing light to the village. Coyote vowed to steal the sun and moon for his own village, only to be told by his own chief, "We do not need these things. They are of no use to us." Nevertheless, Coyote steals the sun and moon, bringing them to the Village of Darkness. Pursuers from the Village of Light give up their chase when they see the Village of Darkness, afraid of the dark. Nevertheless, the chief of the Village of Darkness rejects the sun and the moon, not trusting anything from the other side and viewing both the sun and moon as bad things. However, the people in the Village of Darkness disagree with their chief, embracing the new sources of light so enthusiastically that they make Coyote their new chief.[21]

This tale highlights the role of fear in human reactions to change. People from the Village of Light will not pursue Coyote into the Village of Darkness because they fear the dark, while the Chief from the Village of Darkness repeatedly rejects the new sources of light, both because of their origins and because of their effects, evidencing a profound fear of the new and of change (or perhaps just an overly stubborn adherence to tradition). Nevertheless, when the trickster Coyote brings these changes anyway, the rest of the people in the Village of Darkness embrace them, appreciating the addition of light into their lives. Indeed, they change how their village is governed out of appreciation for Coyote bravely trying new things.

In the Crow tale, Old Man Coyote and Raven worked together long ago—along with Wolf, Bull Moose, Elk Stag, and Buck Antelope—to steal summer from Old Woman, all because Old Man Coyote was continuously cold. Through an elaborate plot, Old Man Coyote steals the black bag with summer in it, then engages in an extended relay race with the other animals to keep Old Woman's children, who are in hot pursuit, from retrieving the bag. When they are safely returned to their own lands, Old Man Coyote opens the bag and releases summer, and the earth rejoices. However, Old Woman's children eventually appear at Old Man Coyote's tipi, demanding that he return summer to them and threatening war. In a tale lawyers should love, Old Man Coyote and the children then negotiate a compromise, whereby each group gets summer for half the year. Thus, the humans in Old Man Coyote's lands now enjoy summer for six months each year.[22]

In this tale, like the Miwok tale, the trickster brings profound changes to both worlds/cultures. Moreover, the result originally posited is total loss or total win—*either* Coyote *or* Old Woman gets summer all the time—just as, in the Miwok tale, *either* the Village of Light *or* the Village of Darkness can have the sun and the moon, but apparently not both. By the end of the Crow tale, however, the two sides have negotiated a compromise that represents a partial win for both, the avoidance of a more destructive war, and real but less profound change for both than might have occurred if the trickster were left to his own devices. Both sides of the fight for summer are active participants in their eventual fate, acting to initiate (Coyote), resist (Old Woman), and eventually mutually adapt to climatic change (everyone). Unlike Raven's ravaging of food from the villages in the first tale, sharing natural resources becomes the best resolution—a resolution reached only because, unlike in the Miwok tale, no one is so afraid of change that they cannot act.

Collectively, trickster tales teach us that we are *not* in complete control, that life involves a certain amount of chaos and unpredictability, and that we must, in a very deep sense, learn to roll with the punches—to celebrate the benefits that can arise from such chaotic interventions as well as deal with the damage that results when change occurs. As importantly, however, trickster tales emphasize that, while the trickster is a frequent and powerful actor on the world and on humanity, humans are profoundly not (or at least not purely) passive victims of the trickster's schemes. Instead, humans have a range of options in how to respond, some of which are more productive than others. Because the trickster often gets tricked himself, trickster narratives also teach that we can act to create our reality, and even when we don't get everything we want, we can still improve upon what our conditions would otherwise be—especially when we do not allow fear of change to become an obstacle of its own.

Trickster narratives thus offer a "radical middle" cultural narrative alternative to both the Manifest Destiny (with its emphasis on control, human exploitation, and technology-as-savior) and the tragedy narrative (with its emphasis on fear). Specifically, trickster narratives simultaneously acknowledge that there are real limitations to humans' abilities to completely control their fates and that humans nevertheless can be effective agents in mitigating or adapting to the changes that they cannot completely control.

Thomas and Patricia Thornton argue that the Raven trickster tales from the Native American tribes of the Pacific Northwest make particularly apt cultural narratives for a climate change era.[23] Characterized by "improvisation in the face of unpredictability,"[24] Raven is both

> a driver of, or respondent to, environmental shifts. Although Raven frequently appears as either the harbinger of or an active agent provoking extraordinary ecological events, they are nonetheless not cast in the rhetoric of crisis. Instead, Raven adapts, innovates, and transforms with Earth's changes, sometimes by relying upon his intimate knowledge of local species, sometimes by cunning and wiles, and sometimes by happenstance as a result of his ulterior manipulations, and, at times, buffoonery. In contrast to the overtly mechanistic cause and effect models that prevail in popular and scientific discourse today, the lessons Raven can and does teach offer a multivalent understanding of the place of human activity in the world. Taken collectively, Raven tales . . . emphasize a moral ecology of mutual dependence, intersubjectivity, survival, resilience, feedbacks, and adaptation in the face of ceaseless and open-ended ecological change.[25]

Raven is therefore "an anthropogenic reflection of humanity as one among many competing, strategizing species."[26] In addition, the Thorntons argue, because Raven operates "as a mutable transcender of conventional boundaries," he "anticipates humanity in the Anthropocene, both as an agent (or 'driver') of change through his appetites and aspirations to control things for his own purposes, and as a resilient respondent to change (through coping, mitigation, adaptation, etc.) when earth systems and their constituent elements prove too powerful, dynamic, and complex to be harnessed for the benefit of one being or species."[27]

Climate change is the trickster of the Anthropocene. We can predict, in a general and most direct sense, what increasing concentrations of greenhouse gases mean for the planet: increasing air temperatures; increasing water temperatures; changes to both air and water currents; changes to dominant weather patterns; freak storms and seasonal anomalies; and so forth. However, pinning down the details of what exactly will happen and when gets a lot trickier, particularly as we project the temporal dimension farther and farther into the future. Moreover, what the collection of direct climate change impacts and their synergistic interactions mean for the "normal" operations of complex planetary systems, from large scale to small scale, is often just an unpredictable surprise. Who knew, for example, that increased melting of Arctic sea ice in the summer leads to particularly severe winters in the East the following year?[28] There is a reason, in other words, that an increasing number of scientists, academics, and journalists refer to climate change as either "climate weirding" or "global weirding."[29] Things are not just "changing"—they are getting strange. And unpredictable. Adopting a trickster cultural narrative would help Americans to shift our perception of our own relationship to this strangeness, increasing our own resilience and chances for productively coping with the Anthropocene.

Resilience Theory

The Anthropocene is an era that will inevitably frustrate those who want to continue to believe that humans are in control of SESs and those who seek to avoid change altogether and maintain the status quo. The trickster offers a new vision, one of flexible resilience in the face of continual and accelerating ecological change. As the Thorntons note, "Raven's mutability, adaptability, unpredictability and resilience, his ability to fly away, take a bird's-eye view, and revise his response to changing planetary conditions always leads to sustainment even in the face of environmental transformations."[30]

Like the trickster cultural narrative, resilience theory also changes how humans view their relationship to nature from earlier scientific frameworks of a "balance of nature" and ecological stationarity. Translated into environmental and natural resources law and policy, resilience theory would allow the United States to reframe these governance tools away from the Manifest Destiny, tragedy, and sustainability narratives to promote a more productive engagement with a world of continual and unpredictable change— change that may in fact constrain future human development options. As Chapter 2 described, the Intergovernmental Panel on Climate Change's Fifth Assessment Report concludes that humanity needs to reconsider what "development goals" can look like in an option-constrained—and in many places under many scenarios, *severely* option-limited—future.[31]

The resilience theory described here is from the Holling school and is a systems theory grounded in the concept of *ecological resilience*. We refer to it as the Holling school of social-ecological resilience because it is now often employed to characterize the dynamics of ecological and *social* systems, known as complex SESs.

There is a lot to unpack in that short description of resilience theory. To begin, the Holling school of resilience theory incorporates the concept of ecological resilience, a different kind of resilience from what we usually have in mind when we use the term "resilience" or "resilient" in everyday speech. Understanding the importance of resilience theory thus requires first teasing out the various important meanings of "resilience."

In the term's most common usage, people generally deem something "resilient" if it can withstand shocks and disturbances or bounce back quickly from those shocks and disturbances to resume a prior form or state of being. Resilience theorists refer to this "bounce back" kind of resilience as *engineering resilience*—the capacity of a system to, essentially, ignore or shrug off a shock or disturbance. Engineering resilience is obviously important to engineering itself, as when architects design skyscrapers in Los Angeles and San Francisco to withstand earthquakes or engineers design river bridges to withstand flooding. However, engineering resilience is also often invoked in connection with people who bounce back from tragedies, communities that come back after a natural disaster, and even ecosystems that recover from a fire or flood.

Engineering resilience is an important concept for ecosystems and SESs. The fact that most complex systems exhibit engineering resilience is one reason why environmental and natural resources laws based on mechanistic models of nature have been able to operate for several decades. Mechanistic models of SESs *can* work—but only on a small scale, over the short term, and under relatively stable ecological conditions.

However, engineering resilience is not the only kind of resilience that natural systems exhibit—the recognition of which is one of the most important contributions of resilience theory. In general, resilience theorists define *ecological resilience* as "the capacity of a system to absorb disturbance and still retain its basic function and structure."[32] In their book *Resilience Practice,* Brian Walker and David Salt define ecological resilience a bit more completely as "the capacity of a system to absorb a spectrum disturbance and reorganize so as to retain essentially the same function, structure, and feedbacks—to have the same identity."[33] In either case, by emphasizing systemic response to disturbance, resilience theorists' focus when discussing ecological resilience is on *change,* underscoring that natural systems *absorb* and *adapt to* change and disturbance as well as resist them and shrug them off. For example, suppose that a factory locates upstream of a freshwater wetland and begins to pipe toxic wastes into that wetland. It will be a rare wetland that can spontaneously eliminate this toxic waste entirely— one reason that we have laws to prevent toxic water pollution in the first place. Nevertheless, to a certain point, most wetlands can sequester many toxics in the submerged soils, fairly effectively cleansing the water itself and isolating the toxics from many (although not all) biological processes. The wetland exhibits ecological resilience—it can absorb the disturbance (at least for a while) without losing its essential structures and functions or its identity as a wetland. However, it has not bounced back into being the same pristine wetland that it used to be.

Moreover, if the factory continues to dump toxic pollution into the wetland, at some point the wetland will exhaust its ability to effectively deal with those toxics, potentially crossing an *ecological threshold* into a different state of being. Ecological thresholds are the boundary or defining conditions of a complex system's current state of existence—e.g., freshwater wetland. When the system crosses (or is pushed across) a threshold, it transforms into something else—in the current example, a toxic swamp that poisons the animals that drink and feed there.

The wetland also illustrates another important element of a system's ecological resilience: Its *capacity for self-organization.* This capacity encompasses the system's development of stabilizing feedbacks among system components that maintain the system—for example, the ecological processes that use wetland currents and plant functions to allow the wetland as a whole to sequester toxics in the soil. Systems that must continually rely on external processes or support to maintain themselves are less ecologically resilient than systems that can remain functional and productive through their own capacities. For example, a farm (itself a complex SES) that requires government subsidies in order to keep going from year to year is less resilient than a one that is profitable without outside assistance. Similarly, ecosystems that require constant management interventions to maintain their existing structure and function are less resilient than those that require little in terms of external controls.

The relative dependency of systems on management intervention is closely related to another element of ecological resilience—a system's *adaptive capacity*. Adaptive capacity reflects a system's flexibility and ability to effectively respond to change and is often reflective of both functional diversity and redundancies within a system. The greater the system's ability to formulate effective responses to change, the more ecologically resilient it is. Thus, the wetland's ability to sequester toxics in the soil is an example of its adaptive capacity. Social systems can also possess adaptive capacity, generally described as the capacity of actors, both individuals and groups, to respond to, create, and shape variability and change in the state of the system.[34] The villagers from the Village of Darkness in the Miwok tale exhibit more adaptive capacity than their chief, while all parties in the Crow tale exhibit their adaptive capacity by negotiating a solution that leads to a new reality for everyone.

Both engineering and ecological resilience describe different aspects of maintaining a system's identity. However, resilience theory describes many more features of SESs and offers a particularly nuanced narrative of natural and SES dynamics, describing a world where complex change has *always* been the dominant feature. Most fundamentally, resilience theory is one of several disciplines that incorporates the concept of system complexity. Like systems theory more generally, the Holling school of resilience theory accepts that many planetary systems, including ecosystems and SESs, operate as self-organizing, complex, adaptive systems. This starting point eliminates a Newtonian mechanistic view of nature—a view that posits that if we could just get the physics and chemistry right in sufficient detail, we could explain and predict without surprises every natural phenomenon that occurs. More generally, mechanistic views— the views of nature currently underlying US environmental law and policy—assume that if you understand each component of a system and the rules for how they can interact, you will understand fairly precisely how the system as a whole works, the same way that you can thoroughly and mathematically explain the dynamics of balls on a pool table through the laws of physics.

Systems theory acknowledges that complex systems exhibit behaviors that no amount of physics or chemistry could have predicted because such systems self-organize in unpredictable ways. Weather systems are a classic example. In 1972, Dr. Edward Lorenz, a research meteorologist, noted that tiny differences in initial conditions in a complex system like weather systems can produce large changes in the resulting weather patterns thousands of miles away.[35] He concluded that the flap of a butterfly's wings in Brazil can affect a tornado in Texas, earning the label "the butterfly effect" for his discovery. The relationship between melting Arctic ice in the summer and winter weather in the American East is another example of the butterfly effect.

Systems theory, in other words, supplants linear conceptions of cause and effect—the idea that A causes B which then causes C, and so on. Instead, it describes relationships among parts that are always interacting with and adapting to each other, leading to descriptions of a world that is always moving and changing, rather than remaining static.[36] As several system theorists have noted, moreover, systems thinking is inherently the realm of narrative, spinning evolving tales of dynamic relationships—although it is questionable whether the tales Western cultures tell themselves about how the world works adequately embrace this complexity.[37]

Complexity theory thus offers a profound reconceptualization of mechanistic views of nature and society. Indeed, "the field of complex systems challenges the notion that by perfectly understanding each component part of a system we will then understand the system as a whole."[38] Instead, the whole really is more than the sum of its parts—a complex forest is something different than trees plus plants plus soil plus water plus birds plus animals plus microorganisms plus insects. This "something different" result is largely the product of the fact that new properties emerge when multiple independent components interact with each other in systems, properties unimaginatively referred to as *emergent properties*.[39] In our forest, emergent properties include the complex cycling of nutrients and water through the system and the intricate assemblages of plants, trees, and detritus that create multiple habitats for diverse species that themselves interact in complex food webs. "Emergent phenomena . . . are generally surprising, and may be extreme."[40] As a result, any given complex system "is far from equilibrium"; instead, "the system shows a complicated mix of ordered and disordered behavior."[41]

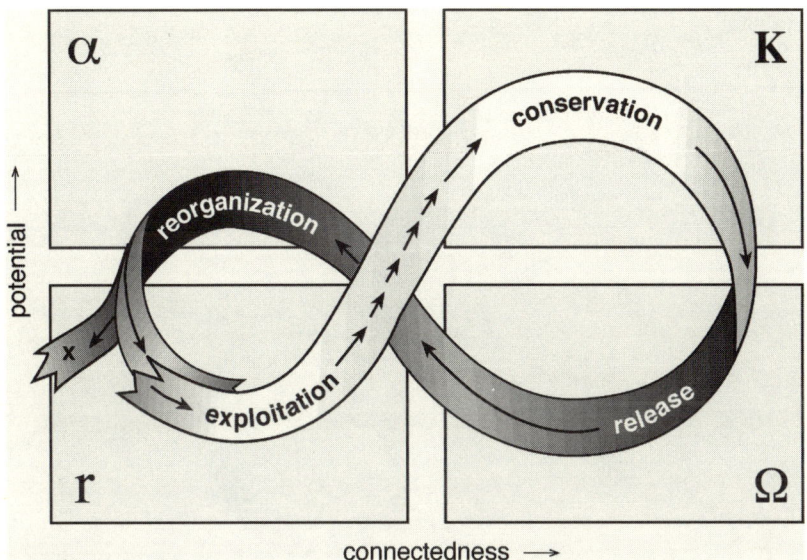

Figure 3.1. The Adaptive Cycle, a representation of four ecosystem functions and the flow of events between them. The arrows depict the speed of the flow within the cycle, with the short arrows representing a slowly changing situation and the long arrows representing a rapidly changing situation. The foreloop (from r to K) is described by uncertainty, novelty, and experimentation and the back loop (from omega to alpha) has a release and a loss of all forms of capital. From Panarchy, *edited by Lance H. Gunderson and C. S. Holling. Copyright © 2002 Island Press. Reproduced by permission of Island Press, Washington, DC.*

In resilience theory, ecosystems, social systems, and SESs are all acknowledged to be complex adaptive systems that exhibit emergent properties. Resilience theory thus reflects the fact that "the last three or four decades have fostered a revolution in the way scientists think about the world: instead of orderly and well behaved, they now view it as complex and uncertain."[42] Again, an important corollary is that natural systems (and SESs) are always changing.[43] To further describe this continual change in ecological systems, in 2002, Lance Gunderson and C. S. "Buzz" Holling described an infinity-loop cycle of change in ecological (and social-ecological) systems, which they termed the *adaptive cycle* (see Figure 3.1).[44] Four distinct phases make up this cycle: growth or exploitation (r); conservation (K); collapse or release (omega); and reorganization (alpha). The adaptive cycle exhibits two major phases (or transitions). The first, often referred to as the foreloop, from r to K, is the slow, incremental phase of growth

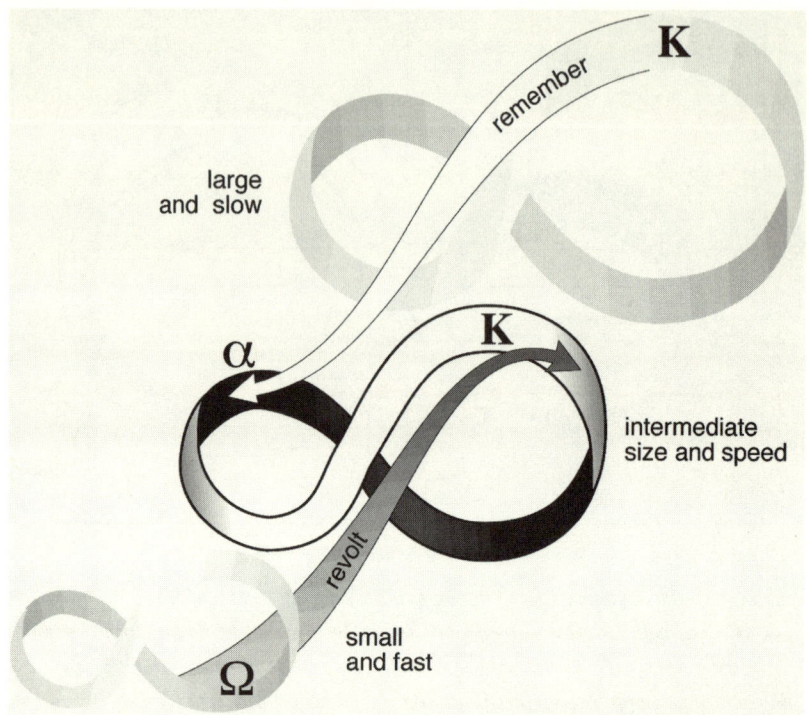

Figure 3.2. Panarchy. Three levels of panarchy illustrated with two connections that are needed to create adaptive capability. The "revolt" connection can cause a critical change in one cycle that leads to an upward flow to a larger and slower cycle. The "remember" connection generates renewal by using the potential that has been created in the larger and slower cycle. From Panarchy, *edited by Lance H. Gunderson and C. S. Holling. Copyright © 2002 Island Press. Reproduced by permission of Island Press, Washington, DC.*

and accumulation. The second, referred to as the back loop, from omega to alpha, is the rapid phase of reorganization leading to renewal. For example, the four phases are exhibited in forest fires: in the conservation phase, the forest reaches maturity and remains relatively stable for decades; when the next fire destroys the large trees, release occurs; and then the system proceeds into reorganization, when colonizing plants and animals potentially compete with the former natives to establish themselves in the former forest.[45]

Complexity further enters resilience theory because adaptive cycles occur at different temporal and spatial scales and that adaptive cycles of change are hierarchically linked, a concept that Gunderson and Holling termed "panarchy" (see Figure 3.2). There are potentially multiple connections between phases. "Two significant connections are labeled 'revolt' and 'remember.' The smaller, faster, nested levels invent, experiment and test, while the larger, slower levels stabilize and conserve accumulated memory of system dynamics. In this way, the slower and larger levels set the conditions within which faster and slower ones function."[46] For example, a forest system "moderates the climate within the stand to narrow the range of temperature variation that the species experience,"[47] and hence stability in the larger forest system (conservation phase) can moderate variations in the larger-scale climate system to promote stability (conservation phase) in smaller sub-systems, such as specific species assemblages in the canopy and on the forest floor. In contrast, rapid change in the larger-scale climate system, such as is resulting from climate change, can drive the small-scale forest across ecological thresholds into new states of being.

Panarchy embodies a systems perspective on natural resources, acknowledging that both social and ecological change resonate in complex ways through multi-scalar SESs. Take, for example, the proposed development of a hydroelectric dam. Even if the legal and policy focal scale is the local community, understanding the social and ecological impacts will require a multi-scalar perspective that looks beyond the community where the dam is located. Ecologically, the dam will change the hydrology of the watershed and may impact native species. At a larger economic scale, regional electricity prices will determine the project's potential for success. At the global scale, climate change will influence the amount of energy that the dam can generate by varying the amount of water available for the project. In addition, these dynamics go both ways. The project will have its own impact on regional electricity prices, and the community's switch to hydropower, a fairly carbon-neutral source of energy, will help to mitigate climate change if it replaces their old coal-fired power plant.

Panarchy alerts policymakers to the complexity of considerations necessary for effective environmental planning and analysis across scales. Panarchy also explains why climate change is such a trickster. In resilience theory terms, there is good reason to believe that the planetary-scale climate system has been in a conservation phase for approximately the last 12,000 years, since the last ice age, a geological period known as the Holocene. Human civilization arose and prospered during this phase of relative climate stability. However, anthropogenic forcing in the form of greenhouse gas emissions is disturbing the climate system. Disturbance at this very-high-level adaptive cycle has consequences for all the linked natural cycles below it—i.e., for every ecosystem and SES on the planet.

As a result, the combination of panarchical hierarchies of adaptive cycles and climate change means that environmental management is an increasingly unpredictable activity. In contrast to mechanistic views of nature, panarchy has always allowed for the possibility that the same management action in a system won't always generate the same response. However, if a larger-scale adaptive cycle (like the climate) is in the conservation phase, its relative steadiness can temper the relative unpredictability of a reorganization phase occurring at a lower scale. Thus, a consistent regional climate can help to ensure that the forest that grows back after a controlled management fire looks a lot like the forest that was burned, which is probably what the managers intended. In contrast, change occurring in a higher-scale adaptive system—like climate change—can make the responses of lower-level systems far less predictable. In the same forest, for example, if the climate system is itself reorganizing, the same prescribed fire may burn out of control, or different species may gain control in the regenerating ecosystem, which may not end up being a forest at all. At the extreme, the panarchical interactions of nested adaptive cycles can overwhelm the managed system's ecological resilience. For example, local coral reef ecosystems are generally ecologically resilient to periodic changes in ocean temperature occurring because of changes at a higher system level, such as during an El Niño or La Niña event. However, in March through May 2016, changes in ocean temperature occurring at the planetary climate-level scale exacerbated ocean temperature increases caused by the 2016 El Niño event. As a result, about 93 percent of the Great Barrier Reef in Australia bleached, a phenomenon that occurs when coral polyps expel their symbiotic (and colorful) algae. Prolonged coral bleaching leads to coral death, and as of November 2016, about two-thirds (67 percent) of the reef had died as a result of this panarchical double-whammy.

Panarchy is a particularly appropriate way of reconceptualizing climate change because the term itself very consciously acknowledges the trickster elements of natural systems and SESs. As noted, in naming these multiscalar dynamics "panarchy," Gunderson and Holling purposefully invoked the Greek god Pan,[48] a trickster associated with unpredictable change. Pan was the god of shepherds and hunters and of the meadows and forests of the mountain wilds. He was known for summoning chaos for those who traversed his realm. The panarchical interactions of nested adaptive cycles similarly describe the complexity and unpredictability of natural systems, revealing an unavoidable element of management chaos that current natural resources law and policy need to acknowledge and incorporate.

As such, resilience theory acknowledges "surprise" and the unpredictable qualities of ecosystems and SESs, as well as that novelty, creativity, and innovation that can occur within SESs. These unpredictable qualities and surprises include changes that are so profound that they shift the identity of the system, generally referred to as *threshold crossings*, *regime shifts*, and *transformations*. For example, the Planetary Boundaries project discussed in Chapter 2 is an attempt to identify, describe, and avoid ecological thresholds at the planetary scale, seeking to avoid regime shifts in planet-scale systems, like the climate system, and hence transformation of the planet as a whole.

The reality that transformation of SESs is always a possibility is an important component of resilience theory,[49] underscoring again the pervasive role of change in this theoretical framework. Perhaps disturbingly to some, resilience theory acknowledges "that the seemingly stable states we see around us in nature and in society, such as woody savannas, democracies, agro-pastoral systems, and nuclear families, can suddenly shift out from underneath us and become something new, with internal controls and aggregate characteristics that are profoundly different from those of the original."[50] Like the trickster cultural narrative, resilience theory acknowledges—this time scientifically—a world of continual ecological change over which humans cannot exercise complete control. Indeed, research in this field indicates that ecological regime shifts have cascading effects that can ripple through social and economic systems as well as ecological, all the while eluding human management strategies that attempt to contain and control them.[51]

One classic example of a regime shift involves eutrophication, a natural process common in aging or polluted lakes or ponds. As the lake gradually builds up its concentration of plant nutrients like nitrogen and phosphorus, these nutrients encourage the growth of algae, plankton, and other plant microorganisms. These microorganisms can take over the pond or lake, forming a blanket of scum that blocks sunlight from reaching the water and prevents the aquatic plants in the system from photosynthesizing, a process that provides oxygen in the water for the other species. Decomposing plants further exhaust the oxygen in the system, leading to a state of hypoxia (low oxygen) or, at the extreme, anoxia (no oxygen, often referred to as a "dead zone"). The original pond or lake system of relatively clean water, vascular plants, and fish and other animal life crosses into a new system state in which the algae dominate, eventually choking the life out of the lake.

Eutrophication is also a good example of a system transformation because, while it is a naturally occurring process, human activities often play a role in initiating or accelerating that process. Humans apply nutrients, like nitrogen and phosphorous, to agricultural lands, suburban lawns, and golf courses in the form of fertilizer. Many of these nutrients eventually run off into lakes and streams, artificially increasing the nutrient load on these waterways, spurring or accelerating eutrophication. Both Lake Erie and the "dead zone" in the Gulf of Mexico are highly publicized examples of eutrophication, and they also illustrate the potentially important role of environmental law in reducing regime shifts. During the 1960s and 1970s, Lake Erie was referred to as a "dead lake." For decades, nutrients from sewage disposal and runoff from heavily developed agricultural and urban lands contaminated the lake and promoted eutrophication. Plant and algae growth choked out most other species living in the lake and made the beaches unusable because of the smell from decaying algae that washed up on the shores. The 1972 federal Clean Water Act imposed pollution controls on sewage treatment plants, leading to drastic reductions in the amount of nutrients entering the lake. Forty years later, while still not totally free of pollutants and nutrients, Lake Erie is again a biologically thriving lake. In contrast, the Clean Water Act has not adequately controlled fertilizer runoff from farms, and agricultural nutrient runoff in the Mississippi River watershed is a substantial cause of the eutrophic dead zone in the Gulf of Mexico, which, on average, covers 6,000 square miles of ocean off the coast of Louisiana and Mississippi, an area the size of Connecticut.

The Lake Erie example suggests that some regime shifts—or at least regime-shifts-in-progress—can be halted and reversed through human efforts. Thus, resilience theory alerts us to the possibility that we can avoid or thwart some system transformations, particularly if direct human actions are the immediate cause of the regime shift (transformation). However, as a result of system complexity, not all such processes are easily reversible. For example, fully eutrophic lakes will not spontaneously shift back to clear water and fish even if all nutrient pollution stops. Climate change also limits reversibility. For example, the Arctic tundra is melting as a result of increasing air temperatures, crossing ecological thresholds and gradually transforming into a different ecosystem that will eventually become a form of shrubland or even boreal forest. Because of time lags in the climate system, there isn't much humans can do to stop this transformation at this point, even if we stop all greenhouse gas emissions tomorrow.

Where regime shifts occur and cannot be reversed, resilience theory emphasizes the system's *transformational capacity*, defined in SES terms as "the system's capacity to reconceptualize and create a fundamentally new system with different characteristics"[52]—the eutrophic lake, the toxic wetland. As we've already seen, when humans recognize their own contributions to this change, they can use law and policy (the Clean Water Act) to slow, halt, and even reverse the transformation—at least in systems where the law actively applies. In addition, even where humans did not cause or cannot fully stop an undesirable transformation, such as the Arctic tundra, they can sometimes act to alter the final destination of that transformation. Thus, resilience theory productively reconceptualizes humanity's relationship with nature in the Anthropocene, suggesting that humans can potentially play a role in guiding some unavoidable system transformations in order to avoid particularly unproductive alternative states of both ecosystems and, more importantly, SESs—especially SESs that have traditionally depended on a natural resource base that is now changing (e.g., logging of forests or commercial fishing). Chapters 4 and 5 in part explore this potential at two different scales (New Mexico forests and global marine fisheries, respectively), demonstrating how resilience theory and resilience thinking allow for continued human agency even when complete human control of a complex SES is not possible.

Going one step further, *intentional transformation* involves a conscious and deliberate negotiation of a system from one system state to another. For example, at the ecological level, Montana's trout streams are becoming too warm to support cold-water species of trout. However, that doesn't mean that the streams have to devolve into dead streams—humans can act to promote a transition to viable warm-water aquatic ecosystems. Similarly, a community that depends on a changing natural resource for its economic vitality can potentially proactively act to change its economic base, productively transforming that SES into a system that can adapt to—even take advantage of—its new natural resource realities. An SES's transformative capacity is defined by the ability of the actors within the system to: (1)

be prepared to change (as opposed to being in a state of denial); (2) have options for change (the identification of possible new "trajectories" for the system shift); and (3) have the capacity to change (the ability to make choices from among the possible new trajectories).[53] Effectively exercising transformational capacity in the Anthropocene, therefore, requires a different cultural narrative about climate change than the four we examined in Chapter 1. It requires acknowledgment both that change is occurring and that human intervention is limited.

Transformative capacity is best viewed as a specific aspect of *adaptive capacity*—the capacity to adapt to and with transformations. As such, adaptive capacity within an SES can serve two purposes within resilience theory. Specifically, adaptive capacity is crucial both when the management orientation is to maintain the current system state and when SES dynamics are such that transformation should or will occur. Elements of adaptive capacity for social institutions include: (1) encouraging the involvement of a variety of perspectives, actors, and solutions; (2) enabling social actors to continually learn and improve their institutions (sometimes referred to as adaptive management and adaptive governance); (3) allowing and motivating social actors to adjust their behavior; (4) mobilizing leadership qualities; (5) mobilizing resources for implementing adaptation measures; and (6) supporting principles of fair governance.[54] Effective law and policy for the Anthropocene must thus keep all of these elements in mind, as well.

Unlike adaptive capacity more generally, transformative capacity highlights an important element of resilience theory that is often overlooked in policy discussions invoking the concept: "Resilience" as a system state is not inherently "good" or "bad." Resilience theory describes system properties rather than setting normative goals for either environmental management or society generally. Indeed, managing to enhance resilience may or may not promote social and environmental goals, depending on whose resilience to what kind of disturbance is being enhanced. There are many examples of relatively stable and resilient SESs that are not desirable, situations in which humans generally want transformation to occur. An algae-ridden eutrophic lake is a stable ecological system, but rarely a desirable system state.[55] Repressive dictatorships can also be remarkably resilient and decisively undesirable, and the "Arab Spring" has been invoked as an example of regime change within a social system.[56]

Any notion of "building resilience" into environmental and natural resources law and policy must therefore be preceded by resolution of three questions necessary to supply the normative goals on which those laws and policies can be based: the resilience *of* what *to* what and *for whom*?[57] In other words, resilience theory can inform management practices by pointing out new ecological realities and challenges, but effectively incorporating that theory into law and policy requires policy and decision makers to first identify overarching systems states (referred to as general resilience) and/or elements of the system (specific resilience) that we want to keep, promote, or transition to. Once the desired outcomes are identified, moreover, decision makers must assess the perturbing factors and disturbances and consider how best to address potential or existing threats to desired ecosystem and SES states in light of the complexities that resilience theory describes.[58]

It is this normative decision-making dimension—the province of governance institutions, law, and policy—that seems to be getting lost in the emerging resilience narrative. We will discuss this point in further detail in Chapter 6. For now, even as major players on the international environmental stage are calling for the greater incorporation of "resilience" into environmental and natural resources law and policy, the normative goals of incorporating resilience concepts tend to be assumed and often take the form of sustainable development. For example, the International Union for the Conservation of Nature (IUCN) has recommended greater use of resilience concepts, noting that "the capacity of nature to meet human needs depends on both its internal dynamics and its dynamic responses to human stresses. The resilience of the biosphere is critical to the sustainability of human enterprise on earth."[59] Similarly, the Intergovernmental Panel on Climate Change (IPCC) has increased its reliance on resilience concepts as a response to climate change, but its normative goal remains sustainable development. Specifically, in its Fifth Assessment Report, it promoted what it calls "climate-resilient pathways."[60] "Climate-resilient pathways are development trajectories that combine adaptation and mitigation to realize the goal of sustainable development. They can be seen as iterative, continually evolving processes for managing change within complex systems."[61] Lacking in the IPCC's discussion, however, are the details necessary to accomplish these tasks. More importantly, neither the IUCN nor the IPCC examines what "sustainable development" can look like in the Anthropocene, nor do they explain how to achieve sustainable development goals in an era where escalating climate change is acting on panarchical, multi-scalar systems, both creating and portending significant transformations in those systems.

The issue for US environmental and natural resources law and policy is substantially more basic. While engineering resilience has been part of their design since the regulatory era of the 1970s, ecological resilience has been largely ignored. Recently, however, the concept of SES resilience from the Holling school has been gaining popularity within natural resource policy in the United States, especially in the planning context.[62] Examples include the US Fish and Wildlife Service's Strategic Plan for Responding to Accelerating Climate Change for the National Wildlife Refuge System,[63] management of National Forest System lands,[64] the US Bureau of Reclamation's Water SMART program,[65] and the National Oceanic and Atmospheric Administration's Next Generation Strategic Plan.[66] As with the international examples, however, these plans and programs invoke the concept of resilience, but there is little to guide actual application of the theory. Importantly, these planning documents do nothing to change the statutory mandates driving natural resource management, mandates that are built on assumptions of stationarity and capacity to control—i.e., humans as engineers. Among these examples, only the National Oceanic and Atmospheric Administration explicitly states an intention to address social as well as ecological resilience.

Planning documents are insufficient to change management paradigms. The mechanistic frameworks from the Manifest Destiny and tragedy narratives would have eventually become a problem regardless, but the Anthropocene and climate change are accelerating the pace at which the laws governing environmental management are becoming disconnected from ecological reality. As such, environmental and natural resources law and policy must embrace ecological resilience and resilience theory to remain practically functional in the twenty-first century. Thus, while the *rhetoric* of resilience is increasingly embraced in US natural resources policy, the actual legal mandates for the relevant agencies are still grounded in past paradigms: US law and policy have yet to fully incorporate resilience *theory*. Chapters 4 and 5 provide examples in the contexts of forest and fisheries management. The next section outlines the more general refinements to US environmental and natural resources law and policy that a more complete consideration of resilience theory would suggest in order to better equip environmental governance to cope with the Anthropocene.

Resilience as Cultural and Legal Narrative

Resilience theory is a new, emerging narrative—one that provides a more helpful orientation toward environmental and SES management in the Anthropocene. It places emphasis on research and policy efforts that help us to understand and cope with change. It allows for a more realistic approach to management in the Anthropocene because it acknowledges continual change and provides a way of thinking about how to foster the SES components and dynamics we value and want to protect.

Resilience theory offers lawmakers and managers a different and more empowering way of thinking about natural resources management. Theoretical tools, including the adaptive cycle, panarchy, and adaptive capacity, are helpful for conceptualizing SESs, but it is important to avoid being overly deterministic or assuming that all systems will behave in a certain way. Resilience scholars have consistently stated that resilience is not a "theory of everything."[67] Scholars including Debra Davidson have noted that the adaptive cycle, in particular, may be overly deterministic when applied to social systems, given the nature of human agency.[68]

Nevertheless, resilience theory offers a fundamentally different narrative of how SESs function than the mechanistic and predictable models of a nature subject to human control that undergird current environmental and natural resources law in the United States. As Chapter 2 discussed, current laws and policies in the United States are based on a number of assumptions that are no longer appropriate. Consider for example, the federal Endangered Species Act (ESA).[69] The ESA is one of the most powerful and controversial environmental laws in the United States—and, given the importance of maintaining biodiversity in the Anthropocene, it is likely to remain a key law for the United States if we choose to prioritize enhancing species' resilience to climate change. Indeed, as a result of its uncompromising position against biodiversity loss, the ESA has become the primary driver of many ecological restoration efforts in the United States.

Unfortunately, however, the ESA currently has limited capacity to effectively engage the complexity of SESs in the Anthropocene. As Chapter 2 noted, the statute seeks "to provide a means whereby the ecosystems upon which endangered species and threatened species depend may be conserved" and "to provide a program for the conservation of such endangered species and threatened species."[70] Thus, the goal of the ESA is recovery of endangered species—to ensure that they are able to survive on their own in the wild. However, the ESA's view of recovery assumes that the protection of a species should be based on the *historic* distribution and range of that species. For example, a species' very status as "endangered" or "threatened" depends on its status in its historical range,[71] and critical habitat protections are keyed to a species' current range.[72] Trickster climate change, however, is currently involved in an "ecological reshuffling" of species and their habitats, making recovery on these terms untenable in most cases.[73] Similarly, the ESA ignores the practical reality, given the level of human manipulation of the natural world, that most threatened and endangered species will never achieve a level of "recovery" that does not require ongoing management.[74] In the Anthropocene, biodiversity protection will require more than protecting certain habitats from human intervention. Active intervention at all levels will be needed to avoid catastrophic losses in biodiversity.[75]

Similarly, broad assumptions of environmental stationarity underlie Congress's policy for the National Forests—namely, "that all forested lands in the National Forest System shall be maintained in appropriate forest cover with species of trees, degree of stocking, rate of growth, and conditions of stand designed to secure the maximum benefits of multiple use sustained yield management in accordance with land management plans."[76] However, as we will see in Chapter 4, the current challenges facing the nation's forests involve bark beetle infestation, wildfire, and other regime-changing forces that make "multiple use and sustained yield" an outdated environmental goal.

The challenge for the Anthropocene is to design new governance structures that thoroughly incorporate resilience theory and do homage to the trickster. To allow both American society overall and individual communities to effectively adapt to change, these legal and policy designs must promote adaptive and transformative capacity and incorporate more administrative flexibility to allow responses to a changing environment to evolve as necessary. At the same time, however, these laws and policies must also provide the necessary strong and enforceable frameworks that will sufficiently support the SES system states that we seek to foster and protect by increasing their ecological resilience to climate change's impacts.

This growing tension between enforceability and flexibility and the challenge of accommodating both within environmental management have become the focus of legal scholars who pay close attention to the interrelationship of conservation science and law.[77] For example, in "General

Design Principles for Resilience and Adaptive Capacity in Legal Systems: Applications to Climate Change Adaptation Law,"[78] J. B. Ruhl provides some suggestions for designing legal systems that are themselves resilient and therefore more responsive to climate change and other challenges. Noting the extent to which this design effort will require a significant departure from the status quo, Ruhl emphasizes how the current legal system is preoccupied with certainty and finality and the difficulty many federal agencies are having in incorporating adaptive management as a primary vehicle for incorporating resilience theory into actual environmental management:

> The problem is that natural resource management agencies are locked in an administrative law system that . . . shows no sign of being flexible in that regard. The [current] system's fixation on predecisional environmental assessment, cost-benefit analysis, records of decisions, and judicial review litigation has only pushed the system toward a "front-end" focus on reliability and efficiency that has made adaptive management exceptionally difficult to implement.[79]

Ruhl focuses on strategies for building adaptive capacity within the legal system. Specifically, he identifies the needs to: (1) move away from the current level of investment in land use planning, NEPA environmental impact analyses, and other processes that are in inherently built on assumptions of stationarity and predictability;[80] (2) embrace strategies that are emerging from new governance theory, which include less emphasis on command and control and more encouragement of collaborative, polycentric, and adaptive models of governance;[81] (3) invoke dynamic federalism as an approach for addressing the multi-scalar dimension of climate change and other challenges;[82] and (4) encourage the formation and maintenance of transgovernmental networks as informal but critical linkages across various scales of governance (local, state, national) that promote information sharing and social learning.[83]

Resilience theory should also prompt more specific legal responses. For example, resilience theory teaches us that anthropogenic impacts like pollution can lower the thresholds for regime shifts in ecosystems and SESs. Thus, one generally applicable strategy for increasing the ecological resilience to climate change of systems we like and want to maintain—assuming that the United States adopts that as a normative goal for at least some systems—is to decrease the other human-induced stressors on those systems. As Chapter 5 will explore in more detail with respect to the oceans, resilience theory counsels that we should be strengthening pollution control laws (e.g., extending the Clean Water Act to nutrient runoff), increasing protections for a variety of habitats and the links between them, and reducing human exploitation of species resources.

Beyond these basics, however, the substantive goals and requirements of environmental and natural resources law for the Anthropocene will likely become—and should become—the subject of intense political and policy debate. We are at a critical juncture with regard to integrating resilience thinking into the structures and rules of governance.[84] *No governance institution internationally or in the United States has yet done the difficult normative work necessary to fully incorporate resilience theory into environmental and natural resources law and policy.* At the international level, for example, both the IPCC and the IUCN have explicitly or implicitly answered the resilience "to what" question for incorporating resilience theory into environmental policy: systems should be resilient to climate change. Neither, however, offers a carefully considered answer to the "of what" or "for whom" questions, with the IUCN noting that we should be enhancing the resilience of the entire biosphere to climate change, while the IPCC seems to be indicating that we should be enhancing the resilience of sustainable development to climate change. Neither offers a particularly concrete answer for formulating specific policy and law. There is, in other words, much policy and legal work left to be done, although the IPCC does emphasize the critical role of climate change mitigation in making climate-resilient pathways to sustainable development possible.[85]

Moreover, social transformations are also likely to be part of our future. As the IPCC's Working Group II summarized in 2014:

> *To promote sustainable development within the context of climate change, climate-resilient pathways may involve significant transformations (high confidence; medium evidence, high agreement).* Transformations in economic, social, technological, and political decisions and actions can enable climate-resilient pathways. Although transformations may be reactive, forced, or induced by random factors, they may also be deliberately created through social and political processes.[86]

The IPCC insists on the importance of transformational capacity, emphasizing that "restricting adaptation responses to incremental changes to existing systems and structures without considering transformational change may increase costs and losses and missed opportunities."[87]

The United States is even farther behind in answering these important normative questions—the questions that must frame what exactly we want our law and policy to be doing in the Anthropocene. Even assuming that we can reach general agreement regarding the "to what?" question—law should promote resilience to climate change—both climate change itself and normal political and economic realities mean that we cannot simultaneously make everything and everyone equally resilient to all aspects of climate change, nor would we want do. Utah and Colorado don't really have to worry about becoming resilient to sea-level rise, and the Great Lakes states don't have to worry quite as much about becoming resilient to disappearing water supplies. But Florida faces both of these threats and more. In other words, climate change is not a monolithic threat, and legal and policy promotions of resilience to it need to be far more nuanced once we get beyond the initial and necessary commitment to climate change mitigation.

Far more difficult, however, is the resilience of the "of what" question, because climate change is increasingly requiring that we make choices among policy and political goals; maintaining the status quo everywhere is becoming increasingly impossible. As even the IPCC admits, all aspects of sustainable development might not be possible in the Anthropocene. If so, do we prioritize ecological viability, social equity, or economic development? The most logical view of these components, as was discussed in Chapter 2, is that ecological viability is a foundational requirement for the other two goals, suggesting that it should receive first priority—but that conclusion is nowhere close to a political consensus in the United States (or most other places, we suspect). As one example, at the planetary scale, this requirement would mean prioritizing the resilience of global biodiversity—one of the Planetary Boundaries project's critical thresholds for the entire planet. More immediately, local SESs may face a seemingly subtle but critically important choice between prioritizing the general economic resilience of a community in the face of threatened changes to local resources, which might be accomplished through social adaptation and transformation, and prioritizing the resilience of the existing economic system, which in turn might require intense investment in significant management of surrounding and embedded ecosystems. The first choice of goals preserves and possibly enhances the general community's economic health (and presumably social welfare) by changing the status quo, but it almost inevitably dislocates particular individuals in the process and hence will often very perceptibly alter the matrix of "winners" and "losers" within the community.[88] The second choice, in contrast, seeks to preserve the status quo and hence minimize immediate disruption, but it does so by engaging in what will in many places become increasingly expensive management that will be increasingly likely to fail, and by perpetuating (and perhaps exacerbating) existing social inequalities.

Resilience theory, properly applied to law and policy, thus opens several cans of potentially toxic political worms: When we say that the resilience narrative is a radical middle path, we aren't kidding about the "radical." Coping with change requires hard work and difficult choices. However, the Anthropocene will make these choices increasingly necessary if Americans want to actively participate in their adaptation to a changing world, rather than sit on the sidelines as the passive victims of those changes. Resilience theory and the trickster offer practical and empowering frameworks for coping with this world.

We have an opportunity to change our story, but in order to do so, we have to let go of previous cultural and legal narratives. To date, there has been a reluctance to do so, mainly because many see shifting the discussion from sustainability to resilience as admitting defeat. But that is not the case. As noted, a resilience orientation in most cases will require more from our laws and institutions because it requires transparent discussions about what we value and the trade-offs we face. Moreover, shifting the governance focus from sustainability to resilience does not require that we abandon the sustainability narrative's goals of *intra-* and *inter-*generational equity; instead, the focus on equity is an important contribution of the sustainability narrative that we should keep. However, the Anthropocene now demands that we think in terms of how to preserve and enhance social equity in the midst of significant change, rather than by shifting priorities and resources within a position of significant stability.

As we noted in Chapter 1, the stories we tell about our situation not only assign meaning to past and current events but also determine the options we perceive as available to us. A resilience narrative has the potential to foster and develop the strategies that are necessary to anticipate and negotiate our complex and rapidly changing world. These strategies will necessarily include a number of elements. One is a healthier relationship to science and other forms of knowledge production in the Anthropocene. When we examine the narratives currently animating most environmental governance regimes, we notice assumptions that science exists to provide regulators and managers with definitive answers about what they should do—how much of a toxic substance can humans and the environmental tolerate? At what concentration of particulates in the atmosphere do human health impacts begin to appear? How many fish of what species can we take out of the ocean and still go fishing for the foreseeable future? However, the scientific method has never been good at (or built for) providing such answers, and a changing planet means that—as in the IPCC reports—science must operate in terms of risk, probabilities, and scenario development. Past narratives reflect a faith in science and technology often based on hubris; the future will require humility and an ability to be comfortable with best guesses.

An additional, related assumption underlying past narratives involves our ability to control situations that are identified as environmental challenges. Unfortunately, the nature and complexity of climate change and other wicked problems do not lend themselves to resolution or control. That is not to say that action is not critical. Rather, future strategies cannot realistically be expected to "solve" these problems. Instead, we will be responding to them as meaningfully and effectively as we can. Moreover, in a panarchical world, future strategies must embrace both formal and informal institutional methods at a variety of temporal and spatial scales.

A narrative of resilience has the ability to embrace these strategies. Rather than relying on science to provide timely answers, resilience thinking focuses on asking interesting questions that allow for refinement of continually emerging understandings of SESs. Rather than identifying an institutional "fix," resilience theory focuses on building adaptive and transformational capacity—a vitally important characteristic whether the goal is to maintain a certain SES state or (as gracefully as possible) transform it into a new one. It is a narrative grounded in understanding and responding to change. It is also a narrative of connection and relationships—humans are not separate from nature but rather part of a complex and dynamic social-ecological system. For these reasons, resilience theory and its cultural narrative analog, the trickster, provide a new way of thinking about our relationship to the environmental and natural resource challenges of the Anthropocene. While we do not deny the very real political and cultural challenges that transitioning to these narratives will create, they nevertheless could provide us a substantially empowering way to reconceptualize social-ecological relations and the stories that we tell ourselves *about* ourselves and our place in the world.

We now turn to two case studies to provide more specific examples of resilience theory in application. As noted, we deliberately chose two very different examples in terms of the scale, system processes, resources being managed, the communities for whom the conceptual framework matters, and the potential breadth of the transformations contemplated. They each illustrate the real, if limited, role of human agency in shaping the way those transformations proceed and highlight various kinds of transformational capacity—as well as the decreasing ability to cling to the status quo. We begin in Chapter 4 with the forests of northern New Mexico, where climate change and other factors are pushing the system past an ecological threshold. We then focus on US ocean policies for living marine resources to assess the challenges of fisheries management in the Anthropocene. Together these two case studies demonstrate how resilience theory and a better appreciation of the climate change trickster can allow us to examine familiar environmental challenges in a new way.

CHAPTER FOUR

Regime Change for New Mexico Watersheds

In Chapter 3, we acknowledged that sustainability and resilience theory can sound a lot alike, particularly when we're talking about the ability of a system to maintain its current structure and function in the face of disturbances and shocks. As that chapter highlighted, however, there are two key differences in resilience theory. First, resilience theory stresses that social-ecological systems (SESs) are always adjusting and changing, making the concepts of ecological resilience, adaptation, and transformation critical to describing those systems accurately and hence to managing them more effectively. Second, unlike the sustainability paradigm, a resilience orientation acknowledges that maintaining an existing system in its current form is not always an option. Regime changes—shifts from one system state to another with fundamentally different characteristics—can and do happen.

To underscore the importance of acknowledging system transformations in the Anthropocene, this chapter provides an example of an SES experiencing regime change: the Rio Grande watershed in northern New Mexico. Figure 4.1 provides a map of the watershed, including the Middle Rio Grande (MRG), Rio Chama, and their forested tributaries, an area with roughly 1.7 million acres of ponderosa pine and mixed conifer forest. Much of the forest headwaters are federally owned and managed by the US Forest Service, with the remainder a mixture of state, private, and tribal lands.

Human occupation of the Rio Grande basin in New Mexico has a rich and complex history. Several Native American Pueblo communities live in the area and have their own extensive history of water use.[1] Ancestral

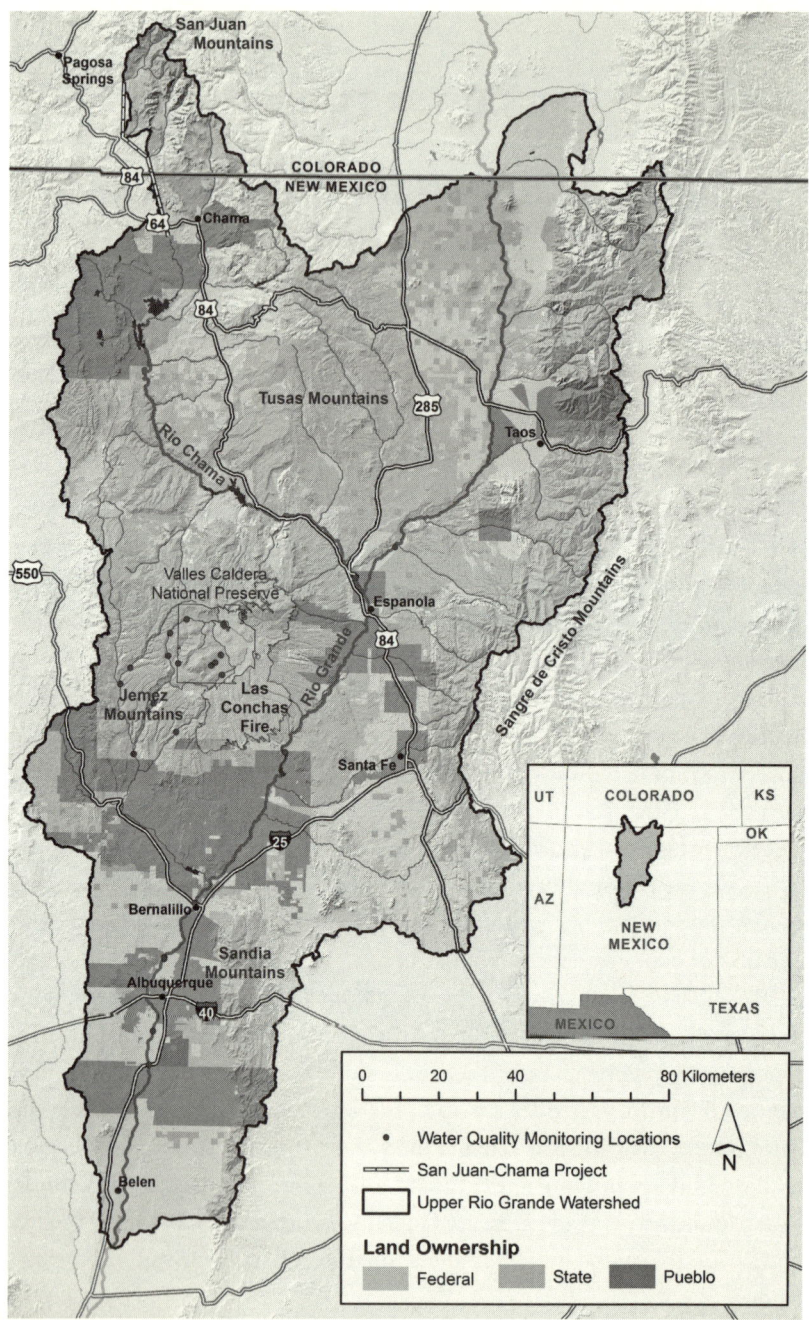

Figure 4.1. Map of MRG Watershed. This map shows the Upper and Middle Rio Grande watershed from the Colorado–New Mexico border to Elephant Butte Reservoir in southern New Mexico, including the Rio Chama watershed from its headwaters in the San Juan Mountains, Colorado, to its confluence with the Rio Grande in Española, New Mexico.

Puebloans underwent several significant migrations, plagued by prolonged droughts that disseminated their subsistence corn-based agricultural system. They brought with them to the Rio Grande Basin sophisticated irrigation practices and adapted these strategies to take advantage of monsoonal rain patterns and annual spring flooding events for subsistence agriculture within the floodplain. Today, the basin is still home to several indigenous Pueblo communities.

In the mid-1500s, the Spanish Conquistadors brought the first wave of European colonialism. Along with them came Spanish settlements in the form of agricultural communities that were granted large tracts of land by the Spanish Crown.[2] These communities established communally governed irrigation systems called *acequias,* resulting in diversions of water from a public waterway and communal division of that water among *acequia* members.[3] For a brief period, the area became part of Mexico, but the United States gained possession of this territory with the signing of the Treaty of Guadalupe Hidalgo in 1848.[4]

This area's incorporation into the United States led to another wave of Euro-American colonialism—this time spearheaded by Anglo ranchers and farmers arriving on the newly constructed railroad.[5] Anglo settlement in New Mexico followed a pattern familiar across the American West, including the encouragement of settlement by various federal Homestead Acts and federal water projects supporting the development of large-scale irrigated agriculture.[6]

Today, the water needs of all these residents, including Pueblo people, Spanish, Mexican, and Anglo settlers, continue to grow. The forests are critical to meeting the needs of these communities, in terms of providing both water supply and water storage. In terms of supply, Albuquerque, Santa Fe, and other communities withdraw water from the Rio Grande and its hydrologically connected groundwater for their municipal water supply. In terms of storage, the forests act as natural reservoirs, holding water in the form of snow at high elevations. In the spring, the snow begins to melt, releasing water into various forest tributaries of the Rio Grande throughout the spring and summer. Functioning forested riparian areas tend to hold water longer than degraded ones because the shade and soil integrity provided by the trees hold the moisture in place. This function also plays an important role in water quality as snow melts and migrates to the small headwater streams, because the forest soil acts as a natural water filtration system.

Global climate change is having a tremendous impact on these forest systems. Average temperatures in the Rio Grande Basin between 1971 and 2011 increased by 0.7 degrees Fahrenheit per decade—a rate approximately twice the global average.[7] For reasons that will be explained, these temperature increases alone—regardless of changes in annual precipitation from climate change—are pushing New Mexico forests past an ecological threshold into regime change and transformation. This transformation will have dramatic implications. The Rio Grande and its tributaries provide water to about half of New Mexico's population, including the downstream communities of Albuquerque and Santa Fe and surrounding agricultural areas.[8] Regime change in New Mexico's forests is also creating unprecedented challenges in governance, requiring forest land managers, local communities, and other stakeholders to design and create new management paradigms to address the ecological regime changes as they occur, mitigate the impacts where possible, and make necessary changes in water storage and use.

Thresholds, tipping points, regime shifts, transformation: All of these terms are used to describe a system transitioning from one system state to another. Sometimes the changes are temporary, but others are long lasting and seemingly permanent. Resilience theory provides a way of thinking about transformational change, which is the first step in formulating a societal response to that change. This chapter begins with a discussion of the role of regime change, transformation, and related concepts at play in resilience theory. It then moves on to explain how temperature and other factors are pushing New Mexico's forest systems past an ecological threshold. Using concepts from resilience theory, it then examines the governance challenges relating specifically to water use and supply for communities in New Mexico that rely on forest systems to provide clean, reliable, and affordable water.

Regime Shifts, Thresholds, and Transformation

One of resilience theory's greatest strengths is its recognition that regime shifts can and do occur. A *regime,* as used in this context, refers to the set of characteristics that define the system boundaries and identity. In the case of forests, this often includes elevation, average temperatures, and precipitation, which in turn dictate the system's hydrological function, types of vegetation, and other forms of biodiversity. A regime *shift* occurs when a system crosses some type of threshold beyond those boundaries into an alternate domain or system state. For example, a dramatic change in vegetation type (from grazing practices, logging, or invasive species) can shift the hydrologic function of a system, causing soil erosion and loss of moisture retention that can eventually lead to a total conversion of the landscape. Logging practices in the American Midwest during the early nineteenth century, for example, converted old-growth and mixed conifer forests into agricultural landscapes. Combined with similar conversions of native savanna and prairie ecosystems for agricultural purposes, the ecological consequences for the region were profound.[9] Many current inhabitants are completely unaware that at one time the forests of Michigan and Wisconsin rivaled the old-growth forest systems still found in parts of Oregon and northern California. What is now agricultural land was once a wild forest system.

Where regime shifts occur, they create fundamentally new systems with different characteristics.[10] Sometimes these regime shifts result from intentional human effort, and conversions of native ecosystems—forests, prairies, wetlands, etc.—to farmland and settlements is a prominent example of intentionally induced regime shifts that have occurred throughout the United States. Sometimes, however, regime shifts occur as a result of changes in climate, natural events like tornados and hurricanes, or because humans unintentionally push ecosystems across ecological thresholds. As Chapter 3 noted, nutrient-loading from fertilizers and habit fragmentation from agriculture and other human activities are creating conditions that are pushing against our planetary boundaries.

How humans cope with these unintentional changes encapsulates one of the biggest challenges for natural resources management in the Anthropocene. Moreover, while restoration has been the traditional governance remedy for such human-induced unintentional regime shifts under existing laws, resilience theory counsels that in the Anthropocene we should be considering intentional transformation—the conscious and deliberate negotiation from one system state to another, in which SESs reconceptualize themselves.

Recognizing this need to manage novel ecosystems in the future, resilience theorist Brian Walker and his coauthors define *transformability* as the "capacity to create a fundamentally new system when ecological, economic, or social (including political) conditions make the existing system untenable."[11] As noted in Chapter 3, a system's *transformative capacity* is defined by: (1) the degree to which managers of the SES are prepared for a change (as opposed to managers being in a state of denial); (2) the identified options for change (the possible new "trajectories" for the system); and (3) the capacity to change (the ability to make choices from among the possible new trajectories).[12] Human agency has long been recognized as an element of transformational processes.[13] Adaptive capacity and transformative capacity are related attributes within an SES, and these capacities are important both when the management orientation is to maintain the current system state *and* when SES dynamics are such that transformation should or will occur. As will be discussed, the concept of transformation is particularly helpful for systems that are approaching potential ecological thresholds, providing managers with a framework for adapting to impending change.

Transformative capacity also highlights an important element of resilience theory that is often overlooked in policy discussions. Remember that a "resilient" system state is not inherently "good" or "bad." There are many examples of relatively stable and resilient SESs that are not desirable. As we discussed in Chapter 3, any notion of "building resilience" must therefore be followed by the questions—resilience *of* what *to* what and *for whom?*[14] In other words, before creating and implementing resilience theory–based policy and law for managing a particular system, it is necessary to first identify overarching system states and/or elements of the system that we want to maintain, as well as those that we would prefer to lose. Once the desired and undesirable attributes and components are identified, managers can perform an assessment of the perturbing factors and disturbances and assess whether they constitute potential or existing threats, as well their capacity to control those threats.[15] Thus, the emphasis of resilience-based management is on building adaptive capacity rather than maintaining stationarity or restoring systems to a prior state. This kind of resilience-based approach to management is more realistic than traditional approaches to natural resources management, especially in the Anthropocene, because it acknowledges nonlinear change and provides a way of thinking about how to foster the SES components and dynamics we value and want to protect.

A Case Study in Transformation: The Forests of Northern New Mexico

When most people conjure images of New Mexico and the American Southwest, they imagine deserts. While desert ecosystems do dominate much of this landscape, forests in New Mexico are significant, comprising nearly 21 percent (or 16.7 million acres) of the state.[16] In fact, the southernmost extension of the Rocky Mountains resides in New Mexico. The five national forests in the state cover most of the state's mountainous areas and are governed by statutes from an earlier era—the Multiple-Use Sustained-Yield Act of 1960[17] and the National Forest Management Act of 1976.[18] As will be discussed, the above legal regimes reflect the old paradigm of stationarity, and natural resource managers are currently struggling to work within the confines of the acts to manage these forests in the face of global climate change.

Some of these New Mexico forest systems, which are critically important, encompass the headwaters of the Rio Grande Basin in northern New Mexico. As noted, they are a key source of water supply for the downstream cities of Santa Fe and Albuquerque and also store water in the form of snow. The numerous perennial creeks, streams, and rivers, including eventually the Rio Grande River, provide habitat for fish and other aquatic species, while associated wetlands and riparian areas critically support the state's biological diversity. Rivers, wetlands, and riparian areas comprise a very small part of the landscape—only about 1 percent. Yet these places play an essential role sustaining New Mexico's web of life. Eighty percent of all sensitive vertebrate species in New Mexico use riparian or aquatic habitats at some time during their life cycle.[19] Changes in the ecological characteristics in New Mexico's forest headwaters have cascading implications downstream.

And those changes are happening. Three interrelated ecological drivers are largely responsible for the regime change occurring in New Mexico's forest systems. The first is temperature. Increased annual temperature impacts the system in various ways. Higher temperatures create longer growing seasons and therefore greater demands from agricultural users and city residents, who need to water plants both earlier and later in the season. Higher temperatures also dry soils and increase the threat of erosion, requiring more water to meet existing needs. In the forest system, rising annual temperatures stress trees in similar ways, with longer growing seasons and higher rates of erosion. High temperatures also create a phenomenon researchers call Vapor Pressure Deficit (VPD),[20] which is the difference between the amount of moisture in the air and how much moisture the air can hold. The warmer the air, the more water vapor it can potentially hold. Once air becomes saturated, water condenses and forms clouds. The more water vapor in the air, the greater the vapor pressure.

However, when the temperature increases, the corresponding pressure "deficit" creates a situation in which the air is so dry that that it begins sucking moisture out of trees. Anyone who has ever tried to ripen peaches or other fruits and vegetables during a hot dry summer has experienced VPD. If you leave a tomato to ripen on your kitchen counter in Albuquerque or some other hot, dry landscape, it will often get wrinkled and pruned before it has a chance to ripen because the summer temperatures make the air so dry that it takes the moisture from the tomato to redress the VPD.

In northern New Mexico, higher temperatures and associated VPD are placing an incredible stress on trees. Based on tree ring data, researchers estimate that, if climate models are correct, average drought stress by the 2050s will match that of the worst years during the largest mega-droughts in at least 1,000 years.[21] These results foreshadow twenty-first-century changes in forest structures and compositions. "Given the reproductive and dispersal limitations of dominant native tree species, climate driven amplification of forest drought stress and associated disturbance processes can be expected to force many landscapes in the Southwest United States and probably elsewhere towards vegetation-type conversions, with species distributions quite different from those familiar to modern civilization."[22]

Two of the "associated disturbances" referenced in this quotation are the other two drivers of New Mexico forest regime change: bark beetle infestation and catastrophic wildfire. Bark beetles are a natural part of many forest systems, including those in the American Southwest. Climate change, however, has bark beetles playing a new role. Higher-than-average temperatures mean that springs come earlier and summers last longer. These seasonal alterations give bark beetles a longer season to feed on trees already weakened by drought stress. In 2012 alone, more than 177,000 acres of pinyon pine, ponderosa pine, and Douglas fir forest in the American Southwest experienced mortality as a result of one or more species of bark beetle.[23]

Both drought stress from VPD and bark beetle infestation make forest systems more susceptible to the third ecological driver—wildfire. Beetle outbreaks cause large areas of upland forest to be susceptible to wildland fires,[24] with the greatest wildfire risk occurring shortly after the infestation and dropping off thereafter.[25] When combined with the drought conditions currently gripping the American Southwest, feedback loops are created among bark beetle outbreaks, forest stress and die-off from VPD, and forest fires, leading to greater areas of forest mortality.[26] This feedback loop has resulted in a dramatic increase in fire frequency, severity, and size over the past decade,[27] and the threat of fires is expected to increase further as a result of climate shifts in the future.[28] Warmer temperatures also mean that this feedback loop takes place for a longer period of time; the fire season in New Mexico is now two months longer than it was thirty years ago.

Fire has always been an important and necessary element of New Mexico's forests, but it is now playing a new role. Tree ring data tell us that, before the late 1800s, fires moved through the landscape every five to fifteen years as low-intensity ground fires that reduced overall fuel loads. However, the railroad came to northern New Mexico in the 1890s, and with it came unintentional fire suppression in the form of livestock grazing. With a means to get their product to market, ranchers embarked on cattle and sheep grazing on an unprecedented scale. By 1910, by denuding the landscape of vegetation, livestock inadvertently disrupted natural fire migration patterns. Eventually livestock numbers declined, and soon after the US Forest Service began its fire suppression efforts in earnest. Fire suppression allowed new trees to grow where meadows and grasslands once were, resulting in unprecedented tree density and the creation of "ladder fuels"—trees that can carry fire up to the crowns of mature trees. By 1990, the forest achieved a maximum density of biomass.[29]

Prior to human-induced fire suppression, fire was a more frequent but less intense part of the ecosystem. Fires seldom burned an entire landscape, instead creating a mosaic pattern in which patches of conifer forest were complemented by aspen groves, scrub oak, and open meadows. However, the high fuel loads combined with extremely dry conditions caused by sustained drought are resulting in what are often called "mega fires," high-severity burns that kill everything in their paths, including microbes in the soils and other elements of the ecosystem needed for regeneration of new plant and tree species.

The Las Conchas fire of 2011 provides one such example of the combined results of all these factors. The fire started when a tree fell on a power line during a period of prolonged drought. The first day, driven by strong and unpredictable winds, the fire burned 43,000 acres—a rate of about an acre per second. It eventually burned over 156,000 acres, becoming the largest fire ever recorded in the Rio Grande watershed. The fire resulted in mandatory evacuation for the nearby community of Los Alamos and Los

Alamos National Laboratory and impacted several other local communities, including the native Santa Clara Pueblo.

Mega fires leave behind a highly degraded landscape subject to erosion and subsequent flooding events. For example, the Las Conchas fire resulted in extreme flooding events and severe water quality problems, impacting both local and downstream communities. Albuquerque's water utility had to shut down its drinking water supply plant, which takes water directly from the Rio Grande downstream, for several weeks when ash from the fire in the upper watershed overwhelmed the system's filtration capacity.[30]

The Las Conchas fire also demonstrated the importance of timing for disturbing events. In the Rio Grande headwater watershed, the wildfire season of May through July is followed by the monsoon rains from July through September. As a result of this timing, fires late in the fire season can lead to extreme flash flooding, debris slides, severely degraded water quality, and associated negative impacts on the natural and human systems that depend on the river and its tributaries.[31] Postfire peak flows in the Rio Grande following monsoonal events have been shown to be ten- to one-hundred-fold higher than baseline conditions.[32]

In sum, the dynamics and feedback loops associated with bark beetle infestation, drought, and fire are having profound impacts on New Mexico's upland forest system. The upland forest system of the Rio Grande watershed is already undergoing a regime change. As a result, when the forest is gone, it will not regenerate with the same vegetation types, nor is restoration of the prior forest a viable option. Mixed conifer forests may be replaced by more temperature- and drought-tolerant pinyon and juniper forests, which are typically associated with lower elevations. Similarly, after the loss of pinyon/juniper forests, the landscape may no longer be covered by forests at all and will likely regenerate with oak scrub, sagebrush, and grasses. The exact succession of species in the uplands will depend on many factors, including not only precipitation trends but also the nature and extent of human influence on the system through reseeding and other forest restoration efforts.

The challenges facing New Mexico's forests are a dramatic example of what is happening in the Rocky Mountain region more generally. In 2014, the Union of Concerned Scientists reported that tens of millions of trees have died in the Rocky Mountains over the past fifteen years and that climate change will significantly increase the impacts from tree-killing insects, higher temperatures, and drought in the years ahead.[33] The Union also predicts that the combined impacts of these stressors will dramatically reduce the ranges of iconic tree species and fundamentally alter Rocky Mountain forests. Mega fires will follow. The US Geological Survey predicts that, as fires become more frequent and less controllable, they will impair the West's ability to absorb carbon dioxide and hence to slow climate change.[34] The West's forests currently sequester more than 90 million metric tons of carbon dioxide per year, about 5 percent of total US carbon emissions. The potential expansion of fires could increase carbon emissions by up to 56 percent by 2050, given future climate conditions, threatening these forests' ability to sequester carbon dioxide over the next twenty-five years.[35]

While certainly not the only ecological driver influencing New Mexico forests, wildfire is an interesting one from a resilience perspective because it provides an excellent example of rapid, nonlinear change within systems. As noted, all of the above stressors and conditions create a forest system approaching regime change. That change may come slowly, as in the case of tree die-off resulting from VPD, or it could come within hours of a bolt of lightning or a spark from an unattended campfire. Regardless, given their greater intensity and frequency, wildfires and associated disturbances are changing the landscapes. Areas that were once forests will become new ecosystems. This transformation in New Mexico's forests will in turn dramatically alter the watershed as a whole and the human and biological communities that have relied on it for centuries.

New Mexico communities are facing unprecedented water quality and quantity challenges as a result of these changes in the upstream watershed. Climate change has already begun transforming headwater stream systems, and climate change impact projections indicate that the resilience of the region's water resources is in jeopardy.[36] While some climate change scenarios are necessarily vague, observations and projections regarding increased temperatures are clear. David Gutzler, a professor in the University of New Mexico Earth and Planetary Studies Department and one of the lead authors of the Intergovernmental Panel on Climate Change's 2013 Assessment Report, states that "warm spring temperatures are one of the clearest observed climate change signals in North America."[37] Increased temperatures mean that precipitation is more likely to come in the form of rain than snow and that snowmelt will occur earlier in the year. Warming temperatures influence not just when and where snows fall, but also how and when snowmelt finds its way—or doesn't—into streams, rivers, and water storage reservoirs downstream. Warming temperatures are making it more difficult to predict how much water will be available through the spring and summer, when water demands are greatest.[38]

Implications for Governance

The social systems in and dependent upon the New Mexico headwater forests are almost as complex as the ecological ones. Upstream, they include federal land managers, such as the US Forest Service and National Park Service, and several native communities, including the Santa Clara Pueblo and Jicarilla Apache tribe. In addition, there are the downstream water users, including the cities of Albuquerque and Santa Fe, along with various agricultural communities throughout the Rio Grande valley. Both land ownership and water use are governed by laws and regulations at local, state, and federal levels.

This section begins with a basic overview of the current governance approaches for land ownership and water use, which are managed separately. It then examines inherent limitations within these existing management paradigms to address the impending ecological regimes shifts discussed above. The section proceeds to outline some new, emerging governance approaches that hold promise in terms of building adaptive capacity within both the ecological and social systems.

Existing Approaches

Federal land managers play a key role in the Rio Grande watershed. The Forest Service, for example, manages the Santa Fe, Carson, and Cibola National Forests in northern New Mexico, as authorized by the National Forest Management Act and Multiple-Use Sustained-Yield Act. Federal land managers also have to comply with various environmental laws, including the National Environmental Policy Act and Endangered Species Act. As discussed in Chapters 2 and 3, all of these laws were created decades ago and are based on a false set of assumptions regarding ecological stationarity, the relevance of historical baseline conditions, and the pace and scale of change. The National Forest Management Act and Multiple-Use Sustained-Yield Act, for example, both set forth "sustained yield" mandates for activities that include timber harvest, grazing, mining, and recreation.[39] However, as discussed in Chapter 2, we have little idea what we can actually "sustain" through the Anthropocene. Even when climate change and other factors were not an issue, moreover, the Forest Service struggled with its multiple-use sustained-yield paradigm. By 1969, the agency had allowed "clearcutting" of 61 percent of western forests and 50 percent of eastern forests under its "sustained yield" mandate before litigation forced it to rethink its approach.[40]

The National Forest Management Act created a planning process that became the primary basis for Forest Service decision making. In 2015, the Forest Service put in place a new planning rule for implementation of the law. Among other measures, the planning rule invokes the concept of ecological resilience, stating that:

> the aim [of ecological restoration] is to reestablish and retain ecological resilience of National Forest System lands and associated resources to achieve sustainable management and provide a broad range of ecosystem services. Healthy, resilient landscapes will have greater capacity to survive natural disturbances and large scale threats to sustainability, especially under changing and uncertain future environmental conditions, such as those driven by climate change and increasing human uses.[41]

While this language sounds encouraging, the planning process is inherently cumbersome. Plans typically take several years to develop and are then revised every fifteen years.[42] Many current forest plans are sorely outdated. The Santa Fe National Forest's current plan was written in 1987, has been amended several times, and is currently in revision. The revised plan for this forest is not due until 2018 at the earliest.[43] As will be discussed, the Forest Service does have several projects to address the increasing fire risks in development in the interim, even under its outdated management plan. However, planning requirements and other regulatory processes, combined with lack of adequate funding, unquestionably hamper the Forest Service's ability to act as quickly and decisively as necessary to address the ecological challenges ahead.

The agency's budget challenges come largely from its attempts to control and suppress wildfires. Fire suppression has expanded from 13 percent to more than 40 percent of the Forest Service's total budget.[44] As a result, the agency has run out of wildfire suppression funds repeatedly over the last decade, resulting in it "raiding" other accounts within the Forest Service to make up the shortfall.[45]

Other landowners also play an important role, including the Valles Caldera National Preserve and Bandelier National Monument, both of which are managed by the National Park Service. In addition to Native American communities, there are some state lands and private landowners in the form of small inholdings and ranches. The intermixed land ownership and multi-jurisdictional nature of the landscape creates part of the management challenge. These issues will be addressed in the section on emerging government paradigms. Of importance here is that each landowner has its own set of responsibilities and limitations. These legal distinctions create an interlocking, multi-jurisdictional landscape that lacks a common land management strategy at a time when the major threats to the landscape—drought, bark beetle, and wildfire—fail to recognize artificial human boundaries.

One of the major problems is referenced above: Land and water are managed separately. Most of the human water use in the system occurs downstream in the watershed. Water is used primarily by agriculture (84 percent). Within the agricultural sector, the main commodities are milk (49.15 percent), cattle (29.8 percent), and alfalfa (5.65 percent).[46] Municipalities, including Albuquerque and Santa Fe, comprise a relatively small portion of water use in the state (8 percent).[47]

Like most western states, water allocation in New Mexico is governed by the prior appropriation doctrine.[48] Prior appropriation is a historically based allocation system that anticipates scarcity. The doctrine of prior appropriation mandates that when shortages occur, the right to use water is determined by the chronological order in which the water was put to beneficial use. "Senior" appropriators are served first. In a water-short year, "junior" appropriators may receive a reduced amount or no water, depending on the supply. Water rights are usufructuary (a right of use as opposed to ownership of the water itself), and water rights must be applied to a beneficial use, broadly defined to include agriculture, municipalities, industry, and fish and wildlife.[49]

When New Mexico formally established the prior appropriation doctrine in 1891, there was no recognition of the values associated with leaving water instream for wildlife and other uses. Until recently, leaving water instream for fish and wildlife was not recognized as a beneficial use.[50] Instream flow rights remain relatively limited and, to date, have been held only on a temporary leasing basis.

In addition, water rights are subject to forfeiture and can be lost if not continually used. The fear of losing a water right as a result of nonuse creates a general disincentive for conservation strategies.[51] Today, the majority of water use supports crop agriculture in a traditional system of gravity-fed flood irrigation.[52] The largest agricultural group is the Middle Rio Grande Conservancy District, which provides irrigation water for about 53,000 acres of crops, primarily alfalfa, and also supports a sizable local dairy industry.[53] Agriculture within the district is mainly small-scale and family owned by Anglos or associated with the six native pueblo communities that are members of the irrigation district.[54]

Like much of the American West, this part of New Mexico has seen a steady increase in population growth and, with that growth, an increasing municipal and industrial water demand.[55] Several municipalities in the area purchase water rights from farmers to meet their growing needs. As a result, there has been a shift in many water rights from their original agricultural use to municipal use.[56] Often, these transactions involve purchases of senior surface water rights to offset the impacts to the river of municipal groundwater pumping.[57] Between 1982 and 2011, 21,000 acre-feet of water rights were transferred,[58] most of which were transfers of agricultural rights to cities such as Belen, Rio Rancho, Albuquerque, and Santa Fe.[59] The competing demands for a limited water supply in this area were highlighted in the US Bureau of Reclamation's 2011 report, which focused on areas of the western United States where existing water supplies are, or will be, inadequate to meet the water demands of people, cities, farms, and the environment even under normal water supply conditions.[60]

The third main social driver in the watershed involves the complex system of built infrastructure in the form of dams and reservoirs that control the flow of water from the upper watershed to the communities downstream. This infrastructure is necessary for both the agricultural and municipal use of the watershed's historic floodplain. Both water delivery and flood control are dependent on a network of ditches, levees, and dams. Virtually all of the water storage for the system takes the form of on-channel surface reservoirs. Moreover, interstate compact agreements often determine how much water can be stored and where. For example, the Rio Grande Compact between New Mexico and Texas requires New Mexico to deliver water to the more downstream Elephant Butte Reservoir, from which evaporation rates are extremely high. Depending on their location, the evaporative losses over the course of a given year from *all* New Mexico reservoirs can be significant. Additionally, there are constraints on how the reservoirs are managed. Many of the water projects are authorized for water storage or flood control, but not both. There are also legal limits on the total amount of water that can be stored at a given time, regardless of the dam's actual physical capacity. Releases of water are often restricted because of the designated safe channel capacity of the river downstream of the dams.

In sum, the social elements involved in New Mexico's water management are primarily based on inherited legal and institutional strategies and associated built infrastructure from decades long past. Land and water are managed separately. Land-use planning and water allocation are inextricably linked in reality but are only loosely associated in terms of actual governance. Prior appropriation, the main approach to water allocation, dates back to when New Mexico became a state in the late 1800s and is based on historic uses rather than contemporary needs. Conservation strategies by cities and towns are succeeding at reducing municipal demand, but agricultural uses, which comprise the vast majority of water use in New Mexico, have no incentive to conserve. Indeed, the incentives for agriculture are quite the opposite: fear of "forfeiting" water rights encourages users to take their full allocation of water each year, even if it is not needed.

The law and policy governing land management in the headwaters is also outdated. The Forest Service is working with a management plan dating from the 1980s that is authorized by a statute from the 1970s. Fire suppression, the official policy approach to natural and human-caused wildfires alike for decades, is now part of the problem. Thus, while the ecological system is experiencing an ecological regime change, the current social system is grappling with the need for its own shift in structure and paradigm. The next section addresses efforts currently under way to meet the coming challenges.

New and Emerging Governance Strategies

The forest uplands of northern New Mexico provide an example for which building the resilience of the current system to climate change is no longer really an option. Instead, we must focus on the ecological transformation under way. In order to effectively respond to these changes and meet the continuing needs for the services these forests provide, the social system will also need to transform. As noted, transformation is a property of complex SESs that occurs when maintaining the existing system state is untenable. As with resilience, transformation of an SES is neither inherently good nor bad. It might be intentional or not, anticipated or unexpected. Revisiting the three elements of transformation noted above, the social system's transformative capacity is defined by: (1) the degree to which managers of the SES are prepared for a change (as opposed to managers being in a state of denial); (2) the identified options for change (the possible new "trajectories" for the system); and (3) the capacity to change (the ability to make choices from among the possible new trajectories).[61]

As to the first element, it appears that most land and water managers in New Mexico's Rio Grande watershed understand that the forest is crossing an ecological threshold. While there may be disagreements over the reasons why or the degree of its eventual impact, there seems to be a general consensus that aggressive action is needed if communities in the area are going to cope with the challenges ahead. Wildfire, forest die-off, and beetle infestation are interlocking causal drivers that make it difficult to predict the exact "when" and "how" of the forest regime's change, but there is a shared understanding that these factors will eventually shift the system into a new state.

The second element involves whether communities can identify options for change, i.e., new ways of being for the system. In this case, options must include coping with change and building adaptive capacity within the SES. There are three main types of possible options for building this adaptive capacity: (1) those strategies focused on upper watershed land management; (2) alternative water use and efficiency strategies; and (3) strategies based on increased capacity for water storage and delivery. Taking each in turn, large-scale forest management cannot keep the forest regime change from happening, but it *can* mitigate the impacts. Drought, climate change, and bark beetles are transforming the system from a pine-dominated landscape to one dominated by other types of vegetation, including scrub oak and grasses. That new dominant vegetation can be managed—if managers are willing and able to have transparent discussions with the public regarding the changes to come. For example, in many parts of the watershed, likely candidates for species succession include scrub oak and aspen. While most managers would agree that aspen is the more desirable species, successful aspen establishments will require stronger, more effective management of species that browse on aspen seedlings and bark. This implicates not only overpopulation of elk in many portions of the watershed, but also federally permitted cattle grazing practices. Changes to these practices might be required to allow the forest transformation desired, to an ecosystem dominated by aspen rather than scrub oak.

With regard to water use and efficiency, rethinking *how* water is allocated could provide significantly larger benefit to the adaptive capacity of the system. There are practical elements of the prior appropriation doctrine that could be used more effectively to create a more realistic system. Currently, water rights are held and irrigation is practiced by individual farmers who select the crops that they grow and the irrigation methods that they use. The result has made New Mexico agriculture the main culprit in water inefficiency. The dominant agricultural crop is alfalfa (a relatively water-consumptive crop) using flood irrigation techniques (a relatively inefficient water delivery strategy). Municipalities, including Albuquerque and Santa Fe, have reduced their per capita water use over the past several years using mainly voluntary strategies. However, the cities have used these savings to accommodate population growth. As a result, these conservation measures generally do not result in less water use. Any true savings in efficiency will require agricultural uses to develop new strategies and incentives for water conservation.

With a different form of water administration, some farmers could be incentivized to grow higher-value crops using more water-efficient irrigation technologies. While prior appropriation creates a general disincentive for water efficiency and conservation strategies, the same doctrine actually prohibits both the "waste" and the unreasonable use of water. Currently, the prohibition against waste is rarely employed, but new adaptive strategies could provide more modern and adaptive definitions of "waste" reflecting current uses and technology-based efficiencies. Water rights could be reviewed periodically—say every five years—to reassess beneficial use and waste.

Several water users in the Rio Grande watershed receive additional surface water from a trans-basin diversion from the Colorado River system. This delivers New Mexico's portion of the river pursuant to the Upper Colorado Compact. Water is diverted from the San Juan River, tributary to the Colorado River, into the Chama River, a major tributary to the upper Rio Grande watershed. Known as the San Juan–Chama Diversion Project, the diversion delivers on average an additional 125 million cubic meters of water into the system each year to water users in New Mexico and provides much-needed adaptive capacity to the system. That said, both the Colorado and Rio Grande watersheds are subject to the ecological drivers of bark beetle, drought, and wildfire. Combined with the fact that the Colorado water flows into the Rio Grande, this means that all of this water is vulnerable to water quantity and quality problems.

Water storage options provide another opportunity to create more adaptive capacity within the system. As already discussed, the Rio Grande watershed is a highly managed hydrological system. The construction and operation of its many dams, reservoirs, and levees may be used to play part of nature's role in water storage. For example, the dams can be operated to release high flows that support the life cycle of native aquatic and riparian species, including endangered species, while also providing water for agricultural and municipal users. The earlier peak runoffs that the system is beginning to experience will require more nuanced and intensive water management, including more management flexibility in the basin's major reservoirs, such as by allowing more storage of water upstream at higher elevations to avoid high evaporative losses at lower elevation reservoirs. Additional management flexibilities could be achieved if water managers are able to rethink the concept of storage to include groundwater storage, storage within the infrastructure of the irrigation network, and temporary storage within functional riparian habitats.

Innovations in underground water storage are already under way. The city of Albuquerque's water utility, for example, recently completed the Bear Canyon Storage Project, its first large-scale effort in aquifer storage and recovery. This project is part of the utility's larger strategy to ensure the aquifer's long-term health and prevent the land-surface subsidence that can happen when aquifers are over-pumped. In 2007, the utility began releasing small amounts of its San Juan–Chama water into the Bear Canyon arroyo (an arroyo is a dry creek bed that fills with water only seasonally or after a heavy monsoon rain) in an attempt to recharge the aquifer. The water authority intends to move from this more passive form of water infiltration to direct injection as well as infiltration to get the water into the aquifer. The goal is to put up to 40,000 acre-feet back into the aquifer for use when surface water is less available.

The city's use of both groundwater and surface water to meet its needs provides an example of what is called "functional redundancy" in resilience theory. It builds adaptive capacity by making the community less reliant on one source of water.

Finally, the third element of transformational capacity involves the ability to *choose* among the identified trajectories and actually start building a new future. There might be a general consensus that forest thinning and riparian restoration would benefit the Rio Grande watershed and enable it to better withstand mega fires and flood events, but is there also a corresponding capacity within the social system to make these things happen? In the land management context, there are several promising initiatives indicating that the answer might be "yes." Steps are being taken toward more effective management and are beginning to address the underlying drivers for increased fire risks and upland forests regime change.

In many respects, the city of Santa Fe is leading the way and providing an example of the types of actions needed by the watershed at large. The city received a grant to work with the Bureau of Reclamation's Basin Study Program, a federal initiative that works with local and state governments to combine scientific information and resource management in order to develop climate adaptation strategies within a specific landscape.[62] In concert with this work, the community developed a twenty-year Santa Fe Municipal Watershed Management Plan in 2013 that authorizes forest thinning, the protocol for water quality and quantity monitoring, and recommends establishing a permanent funding source financed by rate payers for the ongoing protection of the watershed. The city initiated a municipal water user fee to collect funds for watershed protection that has resulted in $7 million in forest treatments.[63]

The city estimates that, based on recent wildfires in the basin, these investments are well worthwhile. "It is estimated that fire suppression and rehabilitation costs associated with a 10,000 to 40,000 acre wildfire impacting some portion of the Municipal Watershed could be between $11.9M and $48M."[64] In addition, "the cost to dredge, haul and dispose of 2,000 acre-feet of sediment and ash from the City's [water storage] reservoirs would likely be between $80M and $240M."[65] The city emphasizes that these costs do not include increased water treatment prices or impacts to the local economy from loss of tourism income. It concludes that, "in comparison to these avoided costs, the cost to treat and maintain forests within the Municipal Watershed is expected to be $5.1 million over 20 years, an average of $258,000 per year."[66]

Another project in the watershed is the South Jemez Mountains Restoration Project, an interagency effort on federal land designed to increase the landscape's resilience to severe wildfire and other large-scale disturbances across the entire upper Jemez River watershed, a major tributary to the Rio Grande.[67] The project, among other things, plans to mechanically thin approximately 30,000 acres and burn about 77,000 acres on the Santa Fe National Forest. These efforts represent the types of action that will be required to delay transformation of the landscape and to help ensure that the transformed ecosystem is as functional as possible. However, in order for work to take place at the pace and scale required, stakeholders will need to make much more of an investment, both financially and institutionally.

One promising effort in this regard is the Rio Grande Water Fund (RGWF). The RGWF is a multi-stakeholder, interagency, and trans-jurisdictional consortium. Led by the Nature Conservancy, the consortium is working to raise the funds necessary for a twenty-year plan to restore roughly 1.7 million acres of overgrown forests around the Rio Grande and its tributaries that remain susceptible to wildfires and the kind of water-quality damage seen in the weeks after the Las Conchas fire.[68] The RGWF's goal is to protect storage, delivery, and quality of Rio Grande water through landscape-scale forest restoration treatments in forested headwater watersheds. The objectives are to: (1) enhance watershed functions by improving the health of streams and riparian areas; (2) mitigate the downstream effects of flooding and debris flows after wildfires; and (3) reduce forest fuels in areas identified as high risk for wildfire and debris flow. The RGWF completed its Comprehensive Plan for Wildfire and Water Source Protection with the endorsement of thirty-seven agencies, stakeholders, and business interests, calling for a tenfold increase in the pace and scale of forest and watershed restoration. Funding would support stream restoration, forest-thinning projects, and postfire rehabilitation covering as much as 600,000 acres in northern New Mexico over the next two decades. At the time of this writing, the RGWF is still seeking the necessary funding to pursue its vision of a more resilient watershed.

The RGWF has been in development since 2012. This collaborative group provides a case study in adaptive governance,[69] and it is both an end-user of the tools developed in adaptive governance research and a subject of study as a model for adaptive management in a region that has no choice but to address drought, climate change, and wildfire. Its vision is of healthy forests and watersheds that provide a reliable supply of high-quality water and other benefits for New Mexico. The challenge becomes how to get the necessary level of investment needed for work at the landscape scale. As Laura McCarthy, the Nature Conservancy's project leader, explains, "We know thinning our forests makes them safer and healthier, but the 3,000 to 5,000 acres we treat each year, on average, is not enough to make a difference."[70] The projected cost of the overall project is $15 million per year over twenty years. Despite the up-front cost, project backers say it makes sense to spend significant amounts of money on these kinds of wildfire prevention measures, noting that thinning an acre of dense forest costs around $700, whereas the economic impact of one acre of scorched land runs as high as $2,125.

One critical element of the overall work in the New Mexico forests is its inter-jurisdictional nature. The RGWF provides one example of how to coordinate the multiple legal jurisdictions and land-use regimes, because that consortium involves agencies at all levels of government, nongovernmental groups, water users, and many others. Post-fire restoration work at the Santa Clara Pueblo north of Santa Fe is another promising example. This Native American community has lived in the Rio Grande watershed for millennia, and it has recently undergone an ecological transformation. The Las Conchas fire destroyed approximately 80 percent of the community's watershed in 2011.[71] As a result, the community (numbering less than 1,000 full-time residents) is now extremely vulnerable to flooding events. For this reason, a study by the Union of Concerned Scientists in 2015 listed the pueblo as one of the top thirty historic sites in the United States most at risk from climate change.[72] The pueblo is located on a floodplain near the Rio Grande River, and water coming down in a nearby canyon can quickly flood the community with a torrent of debris, sediment, and downed trees. "We can't stop the floods, but we can put in different projects to slow it down," said Michael Chavarria, the governor of Santa Clara Pueblo.[73] The pueblo's residents are working with the US Army Corps of Engineers on a series of flood-control structures. Efforts to build adaptive capacity within the system included dredging sediment, installing stream bank protections, and putting in place other immediate flood protection measures. They are also working with the Federal Highway Administration to address damaged bridges and with the US Geological Survey to establish an early warning system to alert residents of rushing floodwaters before they arrive.

In addition, the pueblo is working with adjoining federal landowners, including Bandelier National Monument, the Santa Fe National Forest, and the Valles Caldera National Preserve, to decrease the risk of future wildfire. These types of collaborative, multi-agency, multi-jurisdictional partnerships will be critical for any meaningful watershed-scale work to be successful. Funding will also need to come from a number of sources, including water utilities and other downstream water users, the state of New Mexico, and federal agencies. US senator Martin Heinrich (D-NM) has introduced legislation to promote partnerships that make watersheds more resilient to wildfires and harmful runoff by expanding the existing collaborative forestry program and formalizing Forest Service efforts to remediate and decommission unneeded forest roads. The proposed "Restoring America's Watersheds Act" establishes a Water Source Protection Program within the Forest Service that would bolster nonfederal funding for tree-thinning projects that mitigate wildfire severity and protect water supplies from the harmful effects of ash, sediment, and debris.[74] It would also clarify and enhance the agency's ability to partner with cities, businesses, water utilities, and other water users.

A valuable question to ask when reflecting on all these efforts is whether they are enough. Do they collectively generate the transformative capacity needed to respond effectively to the ecological regime shifts in New Mexico's forests? Only time will tell. Fortunately, all of these efforts may have a window of opportunity in the form of the Pacific Decadal Oscillation (PDO). The PDO is a shift in the temperature pattern of the North Pacific Ocean and occurs on a twenty- to thirty-year cycle, providing relatively high amounts of precipitation to the Southwest, including New Mexico. No one is certain how the PDO works, but there are indications that we are at the beginning of a twenty- to thirty-year cool phase of the PDO.[75] If so, this temporary climate reprieve should lower average temperatures and provide higher-than-average precipitation, allowing those groups already at work in the New Mexico forests a bit of time to reduce fuel loads in these forests and restore riparian areas, which in turn may help to shape the future transformation of New Mexico's forests.[76] The key political and legal question, therefore, becomes whether the social system can adapt and change old governance norms to take advantage of this possible window of opportunity, recognizing the trickster in their midst and taking steps to help guide the region into productive and resilient—albeit transformed—ecosystems.

Conclusion

The projected hydrologic conditions of the American Southwest will stress the Rio Grande watershed in ways that make the current modes of land and water management unviable. New strategies that promote resilience and support system transformation will be needed. The emerging governance strategies described above are promising, but they are not (as of this writing) occurring at the necessary pace and scale. To date, New Mexico has been unwilling to appropriate funds for the RGWF, and currently the city of Albuquerque's water utility, the Albuquerque-Bernalillo Water Utility Authority, has been unwilling to follow Santa Fe's lead and raise water rates to support the effort, although it has created a voluntary donation program.

These examples provide insight into a larger problem—the need to successfully convey the level of urgency involved in this landscape-level challenge. According to a poll conducted by the Water Research Foundation, 92 percent of Americans think water utilities should play a leadership role in helping communities prepare for the impacts of climate change. The challenge becomes taking this general and abstract level of support and connecting it to a specific landscape, which in some cases is located hundreds of miles upstream. Similarly, Max Moritz, a fire specialist at the University of California, Berkeley, notes that people tend to oversimplify the causes of fires, and therefore the solutions.[77] This oversimplification is part of a larger problem when it comes to taking action on climate change. As a general rule, humans do not respond easily to large, slow-moving threats.[78] Even with relatively clear and confident climate projections related to temperature increases and associated regime shifts, it is difficult to collectively complete step one in creating transformational capacity—moving past denial in order to prepare for change.

Resilience theory allows us to recognize that, when a system's current state cannot or no longer should be maintained, it is time to think about building the transformative capacity necessary to support transition to a desirable alternative state. In the case of the Rio Grande watershed, water governance will need to accommodate regime shifts in upland forests that create increasing stress from climate change and drought, and decreasing water availability. The river itself represents the lifeblood of the system, upon which all other social and ecological elements of the system depend. Meeting these challenges will require reconsideration and restructuring of both the built water systems and their institutional structures of governance.

The challenge facing northern New Mexico forests provides an example of how resilience provides a more productive narrative for the future. A tragedy narrative would place great emphasis on isolating the problem and controlling the risk. Fire suppression—the main strategy to deal with wildfire until the 1990s—was reflective of this thinking.[79] While it may have provided some benefits in the short term, fire suppression has now become a part of the problem, with forests loaded with unprecedented amounts of fuel that are more vulnerable than ever to catastrophic fire events. The tragedy narrative operates best in situations where there is relatively low uncertainty and relatively high capacity to control,[80] whereas many of the elements involved here, such as wildfire, are both unpredictable and uncontrollable.[81]

The sustainability narrative also struggles in this context. What can we sustain in this SES moving forward? Efforts to maintain current forest regimes are likely futile, yet the current "multiple-use, sustained-yield" management orientation ignores this reality. The impact of land cover change on water availability is uncertain, and, given the unpredictable and profound impacts of wildfire and drought, any emphasis on sustaining existing resources and rates of associated productivity and growth seem misplaced. The prior appropriation doctrine protects existing water allocations but ignores the reality that many of these allocations were perhaps never realistic historically and are certainly no longer realistic given the projected impacts from climate change.[82]

A resilience narrative embraces the idea that the forest landscape is a dynamic, complex SES that has undergone, and will continue to be characterized by, highly variable rates of change. Rather than trying to determine what can be "sustained," the management emphasis is on understanding the basin's complexity and building its adaptive capacity. Where, as is the case here, the resilience of the ecological system is both weakened and subject to only limited management control, the capacity of the social elements of the system to be adaptive becomes critically important.[83] Without a more adaptive strategy, the social system is unlikely to produce a meaningful and responsive approach to climate change.

CHAPTER FIVE

Marine Fisheries and Biodiversity

How the Trickster Undermines Sustainable Yield

Beginning in late 2013, a persistent "blob" of warm water began forming in the northeastern Pacific Ocean, spreading over two years to affect most of the west coast of Canada and the United States.[1] Death followed, from fin whales and otters in Alaska to starving sea lions in California to dissolving sea stars in tidepools along the entire Pacific coast.[2] As *National Geographic* summarized in September 2016, "Between 2013 and earlier this year, some West Coast waters grew so astonishingly hot that the marine world experienced unprecedented upheaval. Animals showed up in places they'd never been. A toxic bloom of algae, the biggest of its kind on record, shut down California's crab industry for months. Key portions of the food web crashed."[3]

This three-year surge of hot water demonstrated to all that the trickster is alive and well in the Anthropocene's ocean. Panarchical interactions at multiple scales make the ocean a confounding place under the best of circumstances. The ocean is a major driver of the Earth's climate and a modulator of climate change, absorbing much of the heat and carbon dioxide emissions created since the Industrial Revolution. However, it is also subject to multi-scalar fluctuations that can overlap, confound each other, or—as occurred between late 2013 and early 2016—reinforce each other. In the northeastern Pacific Ocean alone, recurring changes include the Pacific Decadal Oscillation, in which "the eastern Pacific flips from a food-rich, cold-water place to something warmer" on a decadal scale (the PDO cycle mentioned at the end of Chapter 4 that might help New Mexico transform); "El Nino, the periodic tropical warming, [which] boosts temperatures in North America" every five to nine years; the California Current, a more constant current that carries "cooler water south from Canada to Baja California"; and numerous upwelling currents that bring cooler and nutrient-rich waters from the ocean bottom to the surface,[4] where they traditionally have helped to support some of the richest fisheries in the world.

Whether climate change had anything to do with the 2013–2016 warming is an open scientific question. Even if it didn't, some scientists view the three-year phenomenon as a "dress rehearsal" of the oceanic changes that are coming.[5] Most importantly for this chapter, all of the ocean's "volatile shifts can redistribute marine life,"[6] and those shifts are becoming both more continuous and more unpredictable. The trickster, in other words, is becoming much more active in our oceans.

While it is easy for humans to view ourselves as separate from the ocean (a manifestation of the tragedy narrative), we are not. Microscopic plants in the ocean known as phytoplankton produce 70 percent of the atmospheric oxygen that we and other land animals breathe; the terrestrial rainforests that receive far more attention generate only 28 percent. Marine fish have also become a critical food source for humans. Of the 7 billion (and counting) people who live on the planet, 1 billion rely on fish (including freshwater fish) as their primary source of protein.[7] As of 2010, 2.9 billion people get about 20 percent of their protein from fish, while 4.3 billion people—well over half the world's human population—get at least 15 percent of their protein from fish.[8] In addition, marine fish are a critical source of protein (30 percent or more of protein consumed) in many specific countries around the world, including several countries in Africa, Japan, Greenland, Taiwan, Indonesia, and several South Pacific island nations.[9]

The ocean thus provisions humans with some of our *most* basic needs—oxygen to breathe and high-protein food to eat. Sheer common sense and self-interest, therefore, counsel that it is in humans' best interest to promote a resilient and productive future for the ocean through the Anthropocene, particularly with regard to marine living resources and marine ecosystems. Policymakers use the term "marine living resources" to distinguish marine species that are valuable to humans from other valuable but nonliving marine resources, such as offshore oil and gas. It is the species and ecosystems, not the minerals, that are of particular concern to marine natural resource law and policy in the Anthropocene because they more intensely both respond to and bear the consequences of changes in the ocean—i.e., they are the primary focus of the trickster's mischief. However, resilience theory counsels that law should pay attention not just to the fish, shellfish, and seaweed that humans consume but also to the more microscopic bases of marine food webs—marine phytoplankton and their animal counterparts, the marine zooplankton—and all the other components of ocean ecosystems.

Chapter 4 reviewed the application of resilience theory to a relatively small, watershed-scale regime shift that is reasonably well understood. This chapter applies resilience theory at a much larger scale—the Earth's ocean. This area of natural resources management by definition operates on much larger scales, and it is currently subject to far greater uncertainty than the New Mexico forests. Contemporary science suggests, however, that we are on the brink of losing significant marine biodiversity, as multiple marine ecosystems not only cross ecological thresholds but also change their compositions and locations.

Indeed, the scale of these changes should obliterate any inclination to cling to our traditional narratives of humans' relationship to the environment. A form of Manifest Destiny prompted explorers to cross and investigate the world's oceans in the fifteenth through nineteenth centuries, eventually developing technologies that allowed humans to exploit living marine resources—whales, fish, seals, turtles, and others—from almost every watery acre. However, available evidence strongly indicates that we have reached the limit of that exploitation. A tragedy narrative is difficult to avoid with respect to the ocean—but the Anthropocene both underscores human dependence on the ocean while simultaneously making clear that the human control component of the tragedy narrative has always been far more attenuated for the ocean than for many other systems. Human ability to even attempt to influence even limited aspects of the ocean is actually decreasing. This is a result of the great uncertainties regarding the ocean's future and of: (1) its large temporal and spatial scale; and (2) the many forces at work in the ocean, from the global carbon cycle (millennial scale) to the global climate cycle (centennial scale) to global and basin-based ocean currents (annual to decadal to centennial scales) that combine to create an extremely complex system. On the social side, ocean governance operates both at the international level (a realization memorialized in the 1958 and 1982 United Nations Conventions on the Law of the Sea) and over multiple generations. Within this reality, humans in fact have only a limited ability to determine the specific future of the ocean, although we have exercised considerable influence over its current changes.

Nevertheless, acknowledging the large-scale planetary and regional cycles that influence the ocean and whose dynamics are largely beyond human control also provides the continuing reason for focusing on marine biodiversity and commercial fishing. Unlike other marine phenomena such as ocean acidification (discussed in more detail below), marine species respond to larger-scale stressors and human management measures at least partially on a human scale, although ecosystem-wide changes can take place over centuries. In other words, if marine law and policy adopts the normative goal to increase the resilience of marine species and marine ecosystems to climate change and ocean acidification, measures targeting species and ecosystems can make an immediate difference. However, to make this difference, we must reframe our management objectives in order to increase the odds that healthy and productive (though probably transformed) marine ecosystems will survive throughout the Anthropocene.

Currently, the laws and governance structures for marine species—particularly harvested species—remain firmly grounded in the sustainability narrative, even though uncertainty thoroughly undermines this narrative. This chapter focuses on the existing sustainability-based marine fisheries management regimes, arguing that such regimes, to the extent that they ever functioned effectively, are no longer suitable in a world of constant change. Specifically, marine fisheries management remains grounded in goals of maximum sustainable yield (MSY), ignoring the reality that calculating what "sustainable" means for harvest of any marine species has become a meaningless endeavor. This is what systemic uncertainty actually means: we have no idea what "sustainable" means for marine species in the Anthropocene because we only very poorly understand the complexities of the multiple drivers, stressors, and disturbances that they face. We do know, however, that changes to marine biodiversity—especially fisheries—are already impacting social systems and their capacity to develop, calling into question marine sustainable development goals as well as MSY-based fish management efforts.

Applying a resilience narrative to the complex ocean system is not easy. One of the first functions of a resilience narrative is to encourage managers to acknowledge that large facets of ocean systems and dynamics are beyond human control. This seemingly disempowering reframing is necessary in order to cast off the vestiges of the tragedy and sustainability approaches. Instead, the resilience narrative suggests—almost demands—that humans adopt a profound humility regarding future human interactions with the ocean in the face of the increasing uncertainties of the Anthropocene. This humility encourages the paradigm-shifting management goal of minimizing human efforts to control marine resources—of getting out of the ocean's way, so to speak. The trickster is profoundly at work in the ocean and, for the time being at least, humans need to just stand back and figure out what he's up to, recognizing simultaneously that we cannot continue to exploit marine species the same way that we have in the past.

The ocean, as a global system, has amazing resilience, and marine ecosystems have demonstrated phoenix-like capacity to reemerge after apparent death. For example, the Coral Castles coral reef ecosystem is located about halfway between Hawai'i and Fiji, part of the Pacific Island nation of Kiribati. At the beginning of the twenty-first century, for poorly understood reasons (like the 2013–2016 warming in the northeastern Pacific), this ecosystem was subjected to unusually warm water, which causes tropical corals to expel their symbiotic algae ("coral bleaching," so named because the algae usually supply most of corals' color) and eventually to die. Indeed, as the *New York Times* reported, "in 2003, researchers declared Coral Castles dead."[10] Nevertheless, in 2008, Coral Castles was included in the Kiribati government's 157,626-square-mile Phoenix Islands Protected Area, protected from vessel traffic and (as of 2015) commercial fishing, but "research dives in 2009 and 2012 had shown little improvement in the coral colonies."[11] In 2015, however, a team of research biologists found Coral Castles once again "teeming with life," which they confirmed in 2016. Randi Rotjan, one of the team members, opined that "if Coral Castles can continue to revive after years of apparent lifelessness, even as water temperatures rise, there might be hope for other reefs with similar damage," but he added that "no one actually knows what drives reef resilience or even what a coral reef looks like as it is rebounding. In remote, hard-to-get-to places, our understanding of coral is roughly akin to a doctor's knowing only what a patient looks like in perfect health and after death."[12] In other words, coral reef science is not refined enough to determine whether Coral Castles "bounced back" (albeit somewhat slowly) as a result of its engineering resilience, adapted to new conditions as a result of its ecological resilience, some combination of the two, or actually transformed into a different kind of coral reef ecosystem that can tolerate warmer temperatures. Regardless, however, the revival of a productive coral ecosystem gives hope that other marine ecosystems can similarly survive the Anthropocene in a productive way. Moreover—although again the science is poorly developed—the fact that Coral Castles revived after receiving significant new legal protections provocatively suggests that "leave it alone" governance measures may indeed help marine ecosystems to cope with the ocean's version of the trickster.[13]

While there is evidence that marine ecosystems and species can be resilient to the escalating numbers of increasingly synergistic threats that they face,[14] their continued resilience to cumulative and synergistic stressors is becoming more and more uncertain, particularly as marine species assemblages shift and intermix in response to the stresses of rising ocean temperatures and increasing ocean acidification. Transformations of at least some marine ecosystems (especially coral reefs and Arctic ecosystems) appear almost inevitable.

For these reasons, resilience theory offers a better paradigm than sustainability for marine biodiversity goals and marine fisheries management under the uncertainties of the Anthropocene. First, as emphasized in other chapters, resilience theory openly acknowledges that regime shifts and ecosystem transformations are part of ecological dynamics and function. Second, the acknowledged potential for regime shifts, in light of the profound uncertainties regarding marine system responses to a multiplicity of stressors, should work to shift our governance orientation from overconfident exploitation to a far more cautious framework and long-term perspective that emphasize governance goals of increasing the adaptive and transformational capacity of marine ecosystems and coastal social-ecological systems (SESs). Such an approach could empower marine fisheries managers in the United States and elsewhere to increase the world's chances of avoiding an impoverished ocean future while simultaneously more deeply acknowledging humanity's vulnerabilities to changes in marine ecosystems. The United States has already demonstrated a willingness to take the international lead in letting vulnerable marine ecosystems simply be, particularly with respect to commercial fishing, establishing a thus far limited but important national capacity to transform marine governance institutions at state, federal, and international levels in ways that both promote marine resilience to changing ocean conditions and give marine ecosystems space to transform without the additional stress of human exploitation. Ideally, changing to a resilience narrative would also aid human climate change adaptation efforts and SES resilience by prompting a significant diversification in the global commercial sources of protein.

The Value of the Ocean

The ocean covers about 71 percent of the world's surface[15] and provides 99 percent of the space available for life.[16] Yet we know very little about the ocean and its ecosystems.[17] Until recently, for example, it was a mystery where Bluefin tuna go when they leave coastal areas.[18]

Nevertheless, we *do* know enough to understand that the ocean is valuable, in terms of both commerce and ecosystem services. In 1997, Robert Costanza and his colleagues estimated that the value of the world's ecosystem services is about $33 trillion[19] and that marine ecosystems contributed 63 percent—almost two-thirds—of that value.[20] In 2005, the Millennium Ecosystem Assessment (MEA) similarly emphasized that "capture fisheries alone [were] worth approximately $81 billion in 2000; aquaculture worth $57 billion in 2000; offshore gas and oil, $132 billion in 1995; marine tourism, much of it in the coast, $161 billion in 1995; and trade and shipping, $155 billion in 1995."[21] While offshore oil and gas development and marine trade and shipping are not dependent on healthy ecosystems, capture fisheries, marine aquaculture, and marine tourism clearly are, meaning that changes to ocean ecosystems can directly impinge on human development and well-being. One well-publicized example clearly illustrates this fact: "The early 1990s collapse of the Newfoundland cod fishery due to overfishing resulted in the loss of tens of thousands of jobs and cost at least $2 billion in income support and retraining."[22]

As we will discuss further in Chapter 6, these ecosystem services determinations only begin to describe the importance of ocean systems. Monetary estimates can be helpful in order to convey the magnitude of what is at stake, but they often fail to capture other valuable elements of the system, including biodiversity. The ocean is a particularly deep reservoir of biodiversity. Indeed, forty-three of the approximately seventy recognized phyla of life—the second-most general classification of life after "kingdom," and hence representing broad diversity—are found in the ocean.[23] Moreover, 45 percent of the known phyla exist *only* in the ocean, and 90 percent of known classes (the next level of classification) are marine.[24]

In 2010, more than 2,700 scientists from over eighty nations completed the first worldwide Census of Marine Life, delineating a comprehensive baseline of Planet Earth's marine biodiversity for the first time. The Census filled many gaps in our knowledge regarding the species that inhabit the oceans, as Census scientists reported "an unanticipated riot of species," raising the estimate for the number of known marine species from 230,000 to nearly 250,000—and "the Census still could not reliably estimate the total number of species, the kinds of life, known and unknown, in the ocean."[25] The Census "found living creatures everywhere it looked, even where heat would melt lead, seawater froze to ice, and light and oxygen were absent. It expanded known habitats and ranges in which life is known to exist. It found that in marine habitats, extreme is normal."[26]

Unfortunately, anthropogenic stresses have long threatened the ocean's wealth. As noted, the pervasive damage in the northeastern Pacific in 2013–2016 is probably just a harbinger of the dramatic changes facing marine ecosystems in the Anthropocene. Long-term stressors like habitat destruction and overfishing are combining with climate change to generate continual alteration of ocean ecosystems, from the melting of Arctic Ocean ice to shifting species. The following is a brief review of those stressors.

Pre–Climate Change Human Stressors on the Ocean, Its Ecosystems, and Its Biodiversity

Despite the biological richness that it found in the ocean, the Census of Marine Life found signs of decline in both species and the sizes of individuals—declines that occurred fairly quickly, sometimes within a human generation. Perhaps most importantly, it found that phytoplankton, the basis of marine food webs and the source of approximately 70 percent of the world's atmospheric oxygen, have declined since 1899.[27]

Humans have long imposed a number of stressors on the oceans, including pollution, coastal development, and marine tourism and recreation. These threats to marine biodiversity and marine ecosystem function are significant individually, cumulatively, and synergistically with each other and with climate change. Before climate change, overfishing and habitat destruction traditionally ranked as the most debilitating stressors to the marine environment. Declines in species abundance from overfishing can be very fast, while the few recoveries have been fairly slow. Indeed, "evidence shows that most species entering human commerce decline, often sharply," and biodiversity at both the top and the bottom of ocean food webs appears to have decreased significantly.[28] Thus, while this section provides an overview of several pre–climate change anthropogenic stressors to marine ecosystems, it emphasizes commercial fishing.

Coastal Degradation

Coastal development and the attendant destruction of coastal habitat are important general stressors for marine ecosystems.[29] The MEA concluded that coastal ecosystems "in many cases are now undergoing more rapid change than at any time in their history, despite the fact that nearshore marine areas have been transformed throughout the last few centuries," including physical transformations from coastal filling and construction and biological transformations from "declines in abundances of marine organisms such as sea turtles, marine mammals, seabirds, fish, and marine invertebrates. . . . These impacts, together with chronic degradation resulting from land-based and marine pollution, have caused significant ecological changes and an overall decline in many ecosystem services."[30] Nor is such degradation likely to cease any time in the near future. Coastal populations worldwide are expected to increase from a density of approximately 77 people per square kilometer to 115 people per square kilometer by 2025, and already "the coastal areas with the greatest population densities are also those with the most shoreline degradation or alteration."[31]

Coastal development also directly spurs pollution problems, such as the "discharge of untreated sewage into the nearshore waters, resulting in enormous amounts of nutrients spreading into the sea and coastal zones."[32] While coastal sewage is not a particularly large problem in the United States, the ocean is one system, and according to the United Nations Environment Programme (UNEP), coastal discharge of untreated sewage has seen the least progress of any coastal pollution problem and would cost at least US $56 billion per year to adequately address.[33] In the meantime, "around 60% of the wastewater discharged into the Caspian Sea is untreated, in Latin America and the Caribbean the figure is close to 80%, and in large parts of Africa and the Indo-Pacific the proportion is as high as 80–90%," putting not just the marine environment but also human health at risk.[34]

Invasive Species

Invasive marine species can opportunistically exploit coastal ecosystems that are already degraded and destroyed while also causing new and independent stresses. Notably, nonnative species can quite successfully travel to new coasts in ships' ballast water. As many as 10,000 marine species are transported in ballast water every day, including marine-facilitated human diseases such as cholera.[35] Invasive species also escape from aquariums, are unintentionally transported by marine tourists, or are intentionally introduced into new marine ecosystems, such as for aquaculture. In total, researchers have identified fifteen vector categories (categories of ways in which marine species are introduced to new environments) for marine invasions.[36] Moreover, transformed coastal habitats facilitate successful invasions,[37] showing the synergistic impacts of anthropogenic stressors. "Similarly, toxic algal blooms often occur in response to environmental pollution, but their frequency and ubiquity have been enhanced by the distribution of algal species around the world."[38]

The rate at which some marine species successfully change environments is impressive. Scientists estimate that San Francisco Bay becomes home to a new invasive species every thirty-two weeks (about three new species every two years), while new invasive species infest other large ports in the United States, Australia, and New Zealand at least every eighty-five weeks, and that rate appears to be increasing.[39] Once introduced into new environments, invasive species can alter marine ecosystem function and ecosystem services and reduce native biodiversity. "Once an invasive alien species enters the local marine environment, it is most likely there forever. It will interact with existing communities and, in the process[,] modify native habitats. Many invasive species can be considered system engineers— that is, rather than just blend into their new environment, they will change it."[40] The results often directly affect fisheries. For example:

> The Asian clam[,] *Potamocorbula amurensis,* now reaches densities of over 10,000/m^2 in San Francisco Bay, and has been blamed for the collapse of local fisheries. An invasive crab, *Carcinus maenas,* a European species now found in Australia, Japan, South Africa and both coasts of North America, is blamed for the collapse of bivalve fisheries on the North American east coast, and it is feared it will outcompete migratory bird populations on the west coast of North America for favoured shellfish.[41]

Marine Pollution

A variety of sources of marine pollution affect marine biodiversity. As noted, in many parts of the world, sewage discharges remain an important source of coastal pollution. The oceans also suffer from another form of land-based pollution: nutrient runoff. Water flowing over and from farms, in the forms of both irrigation return flows and runoff from rain or snowmelt, carries excess fertilizer (mostly nitrogen compounds) to the ocean.[42]

Nutrients also reach the waters through atmospheric deposition, such as from the burning of fossil fuels.[43] Once there, nutrients induce large blooms of marine plants—phytoplankton and algae—often aided by invasive species. As the blooms die off, their decomposition consumes most to all of the oxygen in the water column, leading to hypoxic (low-oxygen) or anoxic (no-oxygen) conditions that make large areas of the ocean uninhabitable by marine animals.[44] In the United States, the largest of these so-called "dead zones" occurs seasonally in the northern Gulf of Mexico at the mouth of the Mississippi River and can reach the size of New Jersey—over 7,000 square miles.[45]

Dead zones are now common throughout the world's coastal regions, often impinging on fisheries.[46] The number of dead zones in the world's seas has doubled every decade since 1960 as a result of increasing marine pollution, and a 2008 study identified more than 400 dead zones throughout the world.[47] The world's biggest dead zone is in the Baltic Sea, where sewage and nitrogen fallout from the burning of fossil fuels combine with fertilizer runoff to over-enrich this small, contained marine environment.[48] Perhaps most disturbingly, dead zones are missing biomass compared to what would be expected, suggesting that the oxygen deprivation can have long-term effects on the region's biodiversity and productivity.[49]

Plastic pollution is causing its own problems. Most plastic does not biodegrade.[50] Instead, it photodegrades, breaking down into ever-smaller particles in the sun.[51] Floating plastic waste accounts for 80 percent or more of marine debris. Various marine animals can become physically entangled in larger forms of plastic debris, leading to injury, dismemberment, and death. Many marine species also consume plastic trash; plastic bags, it turns out, look a lot like jellyfish, which is a food item for sea turtles and other species, and other marine animals intentionally or accidentally consume plastic trash. Once swallowed, the plastic can both inhibit adequate nutrition by taking up space in the digestive system and directly cause death by choking or through internal damage.

In 1997, sailor Charles Moore discovered what has come to be known as the Great Pacific Garbage Patch,[52] an area (actually, two areas) of the northern Pacific Ocean near Hawai'i where a gyre of ocean currents has collected plastic from throughout the Pacific Rim, concentrating it in an area twice the size of Texas.[53] Large pieces of plastic float on the surface of the water, while the much smaller pieces of photodegraded plastic hover just below the surface, suspended in a plastic soup and easily consumed by fish and seabirds.[54] The estimates of just how much plastic is concentrated in this soup vary, but median estimates are that plastic outnumbers the plankton by a ratio of at least 6:1.[55] While the Great Pacific Garbage Patch is the largest of these concentrations, similar garbage patches have been found in the North Atlantic Ocean and the Indian Ocean, while land-derived trash and plastic are found in all of the world's oceans.[56] A 2011 study reported that at least 9.2 percent of fish in and below the Great Pacific Garbage Patch had plastic debris in their stomachs, and the researchers estimated that fish in the North Pacific are ingesting 12,000 to 24,000 tons of plastic every year.[57]

Toxic pollution is also a substantial impairment to marine biodiversity. The MEA noted, for example, that "the estimated 313,000 containers of low-intermediate emission radioactive waste dumped in the Atlantic and Pacific Oceans since the 1970s pose a significant threat to deep-sea ecosystems should the containers leak, which seems likely over the long term."[58] Toxic chemicals like mercury reach the ocean through a variety of industrial processes discharging wastes into upstream waterways and emitting mercury into the air; the relatively heavy mercury then falls out (atmospheric deposition), often onto waterways and the ocean.[59] Many of these chemicals, like methyl mercury, the organic form of mercury, bioaccumulate in marine organisms, becoming more concentrated the further up the food web a species resides.[60] High-level predators such as tuna, swordfish, shark, and mackerel can end up with mercury concentrations in their bodies that are 10,000 times the ambient water concentration.[61] Mercury contamination is already prevalent in food fish, and in 2011, 79 percent of the US coasts were under fish consumption advisories, including all of the Atlantic, Gulf, Alaskan, and Hawaiian coasts, mostly for mercury.[62]

Overfishing

Overfishing is generally considered the primary threat to marine biodiversity and ecosystem function, especially when fishing methods also destroy habitat, as is true with blast fishing and ocean trawling.[63] Overfishing can also interact synergistically with other stresses, such as marine pollution, to destroy the productivity of a particular marine area.[64]

Evidence indicates both that fishing has been promoting marine ecosystem transformation for a long time and that a long-term perspective on marine ecosystems is necessary to assess their ongoing resilience to anthropogenic stressors. In July 2001, Jeremy B. C. Jackson and his colleagues published a review of historical overfishing in *Science,* emphasizing the historical reductions in marine species' abundance and arguing "that major structural and functional changes due to overfishing occurred worldwide in coastal marine ecosystems over centuries."[65] Historical overfishing continues to have ramifications for the health of marine ecosystems:

> Severe overfishing drives species to ecological extinction because over-fished populations no longer interact significantly with other species in the community. Overfishing and ecological extinction predate and precondition modern ecological investigations and the collapse of marine ecosystems in recent times, raising the possibility that many more marine ecosystems may be vulnerable to collapse in the near future.[66]

These overfishing legacies continue to cause ecological problems in marine ecosystems. In the United States, for example, "vast oyster reefs were once prominent structures in Chesapeake Bay, where they may have filtered the equivalent of the entire water column every 3 days."[67] However, overfishing and destruction of these massive oyster beds have left an enduring legacy of hypoxia, anoxia, and eutrophication in Chesapeake Bay, prompting significant legal cleanup efforts. Unfortunately, "now that oyster reefs are destroyed, the effects of eutrophication, disease, hypoxia, and continued dredging interact to prevent the recovery of oysters and associated communities."[68]

Overfishing impacts marine biodiversity and marine ecosystems in three main ways. First, overfishing impacts the targeted species, often to the point of fisheries collapse, which is generally defined as a 90 percent reduction in the species' abundance. Second, through "bycatch," overfishing reduces the populations of nontarget species incidentally caught in nets or on lines, most of which are thrown back in the water dead or dying. Bycatch can be destructive both to the nontarget species and the ecosystem more generally. For example, in the Bering Sea, pollock fisheries catch between 200 and 1,400 metric tons of salmon sharks and Pacific sleeper sharks every year, and "reductions in salmon sharks and Pacific sleeper sharks in the numbers reported could disrupt the Bering Sea ecosystem in unexpected ways, notably by removing predation pressure from a more effective Pollock predator."[69] Finally, overfishing has resulted in a phenomenon known as "fishing down the food web": As larger, more desirable, and higher trophic species are fished out, fishers shift to smaller and once-less-desirable species.[70] As a result, overfishing of the original target species eventually leads to overfishing of far more species at lower trophic levels, far more pervasively disrupting marine food webs. Such disruptions can alter a marine ecosystem's overall ecological state, and restoration may become impossible—even before the trickster climate change began working on these systems.

Climate Change and Its Trickery

Climate change poses the newest and in many ways most pervasive threat to marine biodiversity. Greenhouse gases in the atmosphere set in motion geophysical and geochemical processes at the planetary and regional scales that are both warming the sea and acidifying it, with consequent direct and indirect impacts on marine life. The most important of climate change's impacts in the ocean are changes in ocean temperature, changes in ocean current patterns, and sea-level rise. Moreover, climate change's "evil twin," ocean acidification, is creating its own stresses to marine ecosystem health.

Changes to Ocean Temperatures

One of the most direct impacts of increasing global average atmospheric temperatures is increasing surface sea temperatures (SSTs) and ocean heat content (OHC).[71] NOAA reported, for example, that "upper-ocean heat content for the last several years has reached values consistently higher than for all prior times in the record, demonstrating the dominant role of the oceans in the Earth's energy budget."[72] Indeed, the ocean currently stores 90 percent of the energy from climate change that has accumulated between 1971 and 2010, compared to 1 percent in the atmosphere.[73]

As the 2013–2016 Pacific warming demonstrated, SSTs in parts of the ocean can vary noticeably from year to year as a result of changes in current patterns, such as El Niño and La Niña events.[74] Nevertheless, the overall trend of SSTs since 1950 is up.[75] Indeed, in 2014, the Intergovernmental Panel on Climate Change (IPCC) indicated that most regions of the ocean have warmed 0.36°C to 0.92°C in the first 75 meters of ocean and that ocean warming probably now extends to the sea floor;[76] scientists have detected temperature increases almost 2 miles below the ocean's surface.[77] The IPCC projected that the ocean will continue to warm and that, under a business-as-usual scenario, temperatures in most parts of the ocean will increase by another 2°C to 13°C by the end of this century.[78]

Changes in ocean temperatures cause temperature-sensitive species to migrate poleward,[79] and such migrations have already been detected. For example, in November 2009, researchers at NOAA reported that about half of the commercially important fish stocks in the western North Atlantic Ocean, such as cod and haddock, had been shifting north in response to rising sea temperatures.[80] Unfortunately, temperature-sensitive species at the poles have nowhere to go.[81]

Thus, as a result of increasing ocean temperatures, marine ecosystems are already changing. A few marine species may go extinct because of temperature-induced changes in their habitat or food supply.[82] More importantly, climate change will have more general impacts on marine biodiversity[83] and on fishing and fish stocks.[84] A July 2010 study concluded that ocean temperature is a major determinant of marine biodiversity and that changes in ocean temperature "may ultimately rearrange the global distribution of life in the ocean."[85]

Changes to Ocean Currents

Temperature changes also affect ocean currents.[86] The science-fiction movie *The Day After Tomorrow* capitalized on projected changes to one of the largest of the ocean currents, known as the Great Ocean Conveyor. This global "pump" depends on the sinking of cold water in the North Atlantic Ocean, which in turn pulls warm water from the tropics up the coast of the eastern United States and across the Atlantic Ocean to Europe.[87] In the fifteen years prior to 2009, cold water in the North Atlantic was not sinking as fast as it used to, leading to speculation that the Great Ocean Conveyor was shutting down.[88] However, the sinking of cold water "resumed vigorously" in the winter of 2008–2009, surprising scientists and underscoring just how complex climate change predictions for the ocean are.[89] Even so, in 2014, the IPCC predicted that this large ocean current system would weaken over the twenty-first century, although its abrupt collapse or transition was unlikely.[90]

Even if the Great Ocean Conveyor remains intact, smaller changes to ocean current patterns could still disrupt marine ecosystems at the local or regional scale. Ocean currents are important to marine biodiversity for a number of reasons, but one of the most important of these is upwellings. Upwellings occur when currents carry water up from the bottom of the ocean to the top, bringing nutrient-rich waters to the surface. Because upwellings are nutrient-rich, they support plankton blooms and high concentrations of marine plants and animals, including commercially important species of fish. Upwellings regularly occur off the coasts of California, Chile, and South Africa, and these highly productive areas of the ocean support 20 percent of global fishery yield.[91] Much of the northwest coast of North America benefits from nutrient-rich upwelling currents that support numerous species of fish—and strong fishing industries—in the northern Pacific Ocean.

However, changes in ocean currents can exacerbate the world's ocean dead zone problem even when land-based nutrient pollution decreases. For example, at the beginning of the twenty-first century, a mysterious dead zone grew off the coasts of Oregon and Washington.[92] This dead zone, which occurs in the middle of a commercially important fishery, has been attributed to climate change—specifically, to changing interactions of wind and offshore currents that prevent the normal dissipation of oxygen-deprived waters.[93] Three other such climate change–related dead zones have been detected, one off the coast of Chile and Peru in South America and one each off the west and east coasts of Africa.[94] As a result, increased pollution control efforts are now insufficient to address marine hypoxia problems. Thus, the marine climate change trickster is already altering the necessary foci of marine governance with respect to particular problems.

Sea-Level Rise

Ocean temperature increases also contribute to sea-level rise. "During the past decade, ocean warming has contributed roughly half of the observed rate of sea-level rise, leaving the other half for ocean-mass increase caused by water exchange with continents, glaciers, and ice sheets."[95] According to the IPCC in 2014, global sea-level rose, on average, 0.19 meters between 1901 and 2010, but the pace is accelerating: "The rate of sea-level rise since the mid-19th century has been larger than the mean rate during the previous two millennia (*high confidence*)."[96]

Recent studies indicate that the Greenland ice sheet and Antarctic ice are melting faster than expected.[97] While the unexpectedly increasing pace of polar and glacier ice melting around the world has made predicting future sea-level rise difficult,[98] the IPCC projects that sea-level rise will continue throughout the twenty-first century, reaching an average rate toward the end of the century, under the business-as-usual scenario, of 8–16 millimeters per year, although with regional variations.[99] Other estimates of sea-level rise range from 6 or 7 inches in the next century to a possibility of 215 feet in the next few centuries,[100] suggesting an equally wide—and difficult to predict—range of potential implications for marine (especially coastal) ecosystems and their services. Nevertheless, in 2014, the IPCC did predict, with very high confidence, that "coastal systems and low-lying areas will increasingly experience submergence, flooding and erosion throughout the 21st century and beyond, due to sea-level rise," which will also impact fisheries: "Climatic and non-climatic drivers affecting coral reefs will erode habitats, increase coastline exposure to waves and storms and degrade environmental features important to fisheries and tourism."[101] Sea-level rise will also have a direct impact on many US communities and SESs, because approximately 40 percent of Americans live on or near coastlines.[102]

Ocean Acidification

Increased carbon dioxide in the atmosphere is a primary cause of climate change, which, as noted, induces several physical and biological changes in the ocean. However, because the ocean is the world's largest carbon sink, carbon dioxide also chemically changes the ocean, a phenomenon independent from (albeit causally linked to) climate change. At the beginning of the twenty-first century, the ocean and land ecosystems (mostly plants) were absorbing about half of the anthropogenic emissions of carbon dioxide[103]—roughly 25 percent by land plants and 25 percent by the ocean.[104] The ocean continues to uptake about 22 million tons of carbon dioxide per day.[105] However, because of increasing climate change impacts on the ocean, it appears to be losing its capacity to act as a carbon sink.[106]

In addition, the ocean's role as a carbon sink comes at a cost to the ocean itself in the form of ocean acidification. The oceans are naturally basic, with a pH of about 8.16, and that pH level has been remarkably stable over geological time.[107] However, since the Industrial Revolution, the average ocean surface water pH has dropped by 0.1 unit.[108] While this may sound like a small change, the pH scale is logarithmic, so that a pH decrease of 0.1 unit means that the oceans have become 30 percent more acidic in the last 250 years. According to NOAA scientists, "At present, ocean chemistry is changing at least 100 times more rapidly than it has changed during the 650,000 years preceding our industrial era."[109] Moreover, the ocean's pH is expected to drop by up to another 0.35 units by the end of the century,[110] causing continued ocean acidification to "an extent and at rates that have not occurred for tens of millions of years."[111]

These changes to the ocean's pH can interfere with the ability of a variety of marine creatures—"coral, sea urchins, starfish, many shellfish, and some plankton"—to form and maintain their calcium carbonate shells.[112] Carbon dioxide chemically reacts in the ocean to form carbonic acid, and the presence of carbonic acid reduces the concentration of carbonate in the water, depriving these organisms of raw materials that they need to grow.[113] Both ocean acidification and its effects on sea life have already been observed; as Peter Ward reports, "in the Arctic Ocean, [waters are] now so acidic that one group of mollusks, the pteropods, which are important in the food chain, are going extinct as their shells dissolve off their backs."[114] Ocean acidification is also likely to impair coral reefs. In particular, corals and their associated calcifying macroalgae are predicted to "calcify 10–50% less relative to pre-industrial rates by the middle of this century," leading to declines in coral reef ecosystems and associated loss of marine habitat and biodiversity.[115]

The impacts of ocean acidification could be tremendous, resulting in loss of commercially important fisheries, locally important fisheries, and coastal protection from storms.[116] As researcher Scott Doney and his colleagues emphasized in 2009, "unless there are dramatic changes in fossil fuel use, projected human-driven ocean acidification over this century will be larger and more rapid than anything affecting sea life for tens of millions of years."[117] The IPCC concluded in 2014 that "marine ecosystems, especially coral reefs and polar ecosystems, are at risk from ocean acidification (*medium to high confidence*). . . . Ocean acidification acts together with other global changes (e.g., warming, progressively lower oxygen levels) and with local changes (e.g., pollution, eutrophication) (*high confidence*), leading to interactive, complex and amplified impacts for species and ecosystems."[118]

The Overall Picture: Synergy, Complexity, and Continual Change

The ocean trickster has become an even more complex fellow than he once was, and the changes he is bringing are generally antithetical to human well-being. Climate change will make many existing stressors to marine ecosystems worse, exacerbating the existing threats to marine biodiversity, marine ecosystems, and marine resources. For example, with regard to marine pollution, mercury methylation and the consequent bioaccumulation of mercury in marine organisms appears to be temperature-dependent. As a result, mercury contamination of fish and marine mammals is likely to increase as ocean temperatures increase in response to climate change.[119] Climate change is also likely to increase outbreaks of marine diseases. Such outbreaks signal that the world's marine resources are already over-stressed and vulnerable, and disease outbreaks are increasing among sea turtles, corals, marine mammals, sea urchins, and marine mollusks.[120] UNEP considers the number of outbreaks of marine disease in the last few decades and the resulting mortalities to be "unprecedented."[121]

Particularly disturbing for the future of the ocean is the fact that climate change impacts, especially increases in SSTs and ocean acidification, are already interacting synergistically to impair the ocean's primary production for food webs. As noted, phytoplankton are critical to marine ecosystems,[122] and chlorophyll provides a measure of plant life in the ocean.[123] According to NOAA, "the downward trend in global chlorophyll observed since 1999 has continued through 2009, with current chlorophyll stocks in the central stratified oceans now approaching record lows since 1997."[124] Chlorophyll, and hence phytoplankton growth, is inversely correlated with temperature changes, meaning that as ocean temperatures increase, phytoplankton growth decreases.[125] Loss of phytoplankton poses a potential threat to both marine food webs and atmospheric oxygen.

Climate change is also already impacting world fisheries. As noted, changes in ocean temperatures cause temperature-sensitive species to migrate poleward and, to a certain extent, deeper;[126] scientists expect marine fish stocks to migrate 30 to 130 kilometers poleward, and 3.5 meters deeper, each decade that climate change continues.[127] In May 2013, William Cheung, Reg Watson, and Daniel Pauly concluded in *Nature* that increasing ocean surface temperatures are already affecting fisheries in fifty-two of the sixty-four Large Marine Ecosystems across the world.[128] Such changes, they concluded, could become particularly devastating for tropical and subtropical coastal fishing communities, many of which are already vulnerable to climate change.[129]

Cheung, Watson, and Pauly's study is one of several indicating that humans' ability to rely on wild-caught ocean fisheries is declining and may disappear. Since the end of World War II, fishing effort has been increasing worldwide.[130] Industrial fishing methods and large factory ships that can process caught fish at sea for market freshness allow fishing in all of the ocean, often for months at a time.[131] Equipment can include longlines that can stretch for 20 to 40 miles[132] and trawl nets big enough to hold thirteen jumbo jets.[133]

As a result, marine scientists and others have repeatedly expressed concern that collapses in a variety of marine fisheries stocks are imminent or already occurring. For example, scientific studies published in 2003 concluded that, compared to historic levels, "industrial fishing had reduced the number of large ocean fish to just 10 percent of their pre-industrial population"[134]—90 percent of the populations of large fish species are simply gone. Boris Worm and his colleagues made world news headlines in 2006 when they predicted in *Science* "the global collapse of all taxa currently fished by the mid-21st century."[135] A 2011 study concluded that species collapses are occurring among all trophic levels of fish species, not just among the large apex predator species that tend to be the direct targets of commercial fishing.[136] In December 2013, the World Bank[137] acknowledged that "supplying fish sustainably—producing it without depleting productive natural resources and without damaging the precious aquatic environment—is a huge challenge."[138] Importantly, it accepted that wild capture fisheries, particularly marine fisheries, no longer respond to market prices because of biological limitations[139]—i.e., regardless of how high the price rises and how much effort fishers expend, there are no more fish to be caught. The UN Food and Agriculture Organization (FAO) noted in 2016 that capture of wild marine fish leveled off in about 1980, despite increased commercial fishing effort,[140] again suggesting that humans have reached the limits of marine fishing. In addition, "the share of fish stocks within biologically sustainable levels decreased from 90 percent in 1974 to 68.6 percent in 2013," and areas like the Mediterranean and Black Seas are experiencing "alarming" reductions in catch.[141] Instead, aquaculture is now supplying an increasing share of fish and seafood worldwide, especially in China.[142]

Moreover, troubles in marine fisheries are already undermining the global community's more general sustainability goals. *BBC News* published a particularly poignant example of the human tragedies that can result from ecosystem decline, tracing how the loss of terrestrial food species and especially freshwater and offshore fisheries has led to increased slavery—especially child slavery—in Somalia, Burma, Cambodia, and Thailand.[143] Fewer fisheries and other food species make it highly labor-intensive to get food, promoting the enslavement of children and others to carry out this task.[144] In 2014, the IPCC directly questioned the sustainability of the ocean's resources under these conditions:

Global marine species redistribution and marine biodiversity reduction in sensitive regions, under climate change, will challenge the sustained provi-

sion of fisheries productivity and other ecosystem services, especially at low latitudes (high confidence). By the mid-21st century, under 2°C global warming relative to pre-industrial temperatures, shifts in the geographical range of marine species will cause species richness and fisheries catch potential to increase, on average, at mid and high latitudes (*high confidence*) and to decrease at tropical latitudes and in semi-enclosed seas ... (*medium confidence*). The progressive expansion of Oxygen Minimum Zones and anoxic "dead zones" in the oceans will further constrain fish habitats (*medium confidence*). Open-ocean net primary production is projected to redistribute and to decrease globally, by 2100, under all RCP scenarios (*medium confidence*). Climate change adds to the threats of overfishing and other non-climatic stressors (*high confidence*).[145]

We have entered a world where multiple drivers and feedbacks are interacting with anthropogenic stressors individually, cumulatively, and synergistically. As a result, impending shifts in system function threaten the marine resources upon which millions of humans depend for food, culture and religion, recreation, and livelihoods. Projected collapses of many species in the next few decades, observed transformations such as the destruction of the Black Sea, and repeatedly predicted transformations in the Arctic Ocean and most coral reef ecosystems counsel that humans should be taking a long-term view of marine ecosystems and a precautionary approach to their continued use of living marine resources and should be engaging in pervasive adaptation strategies to reduce their dependence on the sea. Instead, however, current marine natural resource governance systems remain grounded in assumptions of stationarity and stability, as the next section will explore.

Implications for Governance: The Failed Pursuit of Sustainability for Fisheries

Like other natural resources laws, marine fisheries laws assume the general predictability of fisheries resources—i.e., that we know approximately where particular species range and in approximately what numbers. This assumption is embodied in both international and US law in the goal of "maximum sustainable yield" (MSY), a fairly specific implementation of a sustainability narrative, tempered by a tragedy narrative that seeks to prevent and redress overfishing. MSY-based fisheries management is simply unworkable in the Anthropocene because it assumes ecological conditions that no longer exist (and maybe never did). As result, this embedded narrative of MSY poses a number of governance problems for fisheries management in the Anthropocene as marine species continually change in response to climate change, ocean acidification, and synergistic marine stressors.

Maximum Sustainable Yield-Based Management

Under the third United Nations Convention on the Law of the Sea (UNCLOS III, which the non-party United States generally accepts as customary international law), coastal nations must enact conservation measures for marine resources "designed to maintain or restore populations of harvested species at levels which can produce the maximum sustainable yield, as qualified by relevant environmental and economic factors."[146] However, while coastal states can consider the status of a species in setting catch rates, the coastal state must promote "optimum utilization" of these species, so if it cannot harvest the entire allowable catch itself, it is supposed to allow fishers from other nations to do so.[147] Absent a recognized biological need, therefore, UNCLOS III's fishery provisions encourage full (or over-) utilization of marine species, despite the arguments of many that UNCLOS III encourages ecosystem-based management, marine conservation, and biodiversity protection.[148]

In the United States, the federal fisheries statute is the Magnuson-Stevens Fishery Conservation and Management Act. Under this statute, NOAA manages fisheries in federal waters (more than three miles out to sea[149]) at both the national and regional levels. At the regional level, the act creates eight Regional Fisheries Management Councils (FMCs)[150] with responsibilities to create Fishery Management Plans "for each fishery under its authority that requires conservation and management."[151] At the national level, NOAA Fisheries (also referred to as the National Marine Fisheries Service or NMFS) oversees the regional FMCs[152] by ensuring that the Fishery Management Plans meet several national standards.[153] Under the first national standard, plans "shall prevent overfishing while achieving, on a continuing basis, the optimum yield from each fishery for the United States fishing industry."[154] "Optimum" yield, in turn, "is prescribed on the basis of the *maximum sustainable yield* from the fishery, as reduced by any relevant social, economic, or ecological factor."[155] In contrast, "overfishing" occurs when "a rate or level of fishing mortality . . . jeopardizes the capacity of a fishery to produce the maximum sustainable yield on a continuing basis."[156] Thus, like international law, federal fisheries management in the United States is firmly grounded in MSY, although managers can adjust downward if a species or its ecosystem so requires.

Importantly, MSY is not a just legal term (and hence subject to redefinition by NOAA) but also a scientific term of art from fisheries biology. Congress consciously adopted this scientific term into the Magnuson-Stevens Act,[157] and, consistent with biologists' use of the term, NOAA defines MSY to be "the largest long-term average catch or yield that can be taken from a stock or stock complex under *prevailing* ecological, environmental conditions and fishing technological characteristics (e.g., gear selectivity), and the distribution of catch among fleets."[158]

Even without considering climate change, for both scientific and political reasons, achieving MSY without crossing into overfishing is, as a practical matter, very difficult. For example, even with good science, market forces can continue to drive overfishing, as is clear in tuna fisheries.[159] In addition, MSY calculations are based on individual species and assume that any fish of a given species not needed to reproduce and replace the fished individuals of that species are "surplus," ignoring the fact that those "surplus" fish probably play roles in the relevant ecosystem and food webs. As such, MSY-driven stock assessments based on single species can obscure ecosystem impacts that affect other species and overall ecosystem function—including processes that support the fished species. As one example, according to the FAO, "it is ecologically impossible to harvest all species at the MSY level simultaneously. Therefore, some stocks may need to have their abundance maintained above the MSY level to avoid ecosystem overfishing."[160] As a result, MSY-based fisheries management has not been successful in achieving sustainable fisheries, as the various reports and studies discussed above indicate (and even acknowledging other pervasive problems like illegal fishing).

The Clash between MSY and the Anthropocene

The state of world fisheries suggests that the world's pursuit of sustainable fisheries has generally been unsuccessful, despite what may be called local improvements and successes. However, MSY-based fisheries management is a particularly bad fit for the Anthropocene and an ocean of constant change. Resilience theory counsels that a multi-scalar, dynamic, and system-based framework is necessary for successful natural resources management, *especially* in the Anthropocene, but MSY-based laws make it exceedingly difficult for managers to consider system dynamics in a constantly changing world. While optimum yield calculations under the Magnuson-Stevens Act can clearly take account of ecosystem dynamics, the MSY calculations on which they are based focus on specific stocks,[161] perpetuating a data gap and translation problem between the MSY and any attempt at ecosystem-focused optimum yield calculations.

In addition, fishing to MSY *by definition* reduces the overall populations of the target species from their natural maximums and, as noted, assumes that fish are "surplus" if they are not needed for replacement breeding.[162] As a result, in addition to ignoring what the "surplus" fish might be doing in and for the dynamics of the larger ecosystem, MSY figures the natural population of a given fish stock as, essentially, too large. The resulting systemic reduction in fish populations is also exacerbated because fishers target the largest members of the species, which tend to be the most prolific breeders. Fishing therefore disproportionately reduces the breeding capacity of many species. In terms of ecological resilience, the pursuit of MSY reduces targeted species' resilience to other kinds of shocks and disturbances in the system by reducing both response diversity (the lack of prolific breeders) and the species' simple numeric chances of survival.

MSY-based fishing also disturbs the panarchical interactions of nested marine ecosystems in ways that are, at best, poorly understood. However, the historical studies discussed above strongly suggest that marine fishing has been undermining both the engineering and ecological resilience of marine ecosystems for decades to centuries, making ecological thresholds easier to cross even before the complex and synergistic workings of climate change and other anthropogenic stressors came into play.

Most damningly, however, the calculation of MSY assumes constant environmental conditions. As NOAA's definition emphasizes, MSY is calculated "under prevailing ecological, environmental conditions" and reflects a long-term average of sustainable take under those conditions.[163] However, if marine environments are constantly changing, there is no scientifically valid way to calculate MSY because there is no way to know how the targeted population will respond on a long-term basis. In short, MSY-based management simply cannot, by definition, cope with the ocean trickster in the Anthropocene.

And yet, that is exactly what fisheries managers are currently trying to do, a significant mismatch between natural resource governance and social-ecological reality. Climate change and ocean acidification in particular are actively rearranging marine ecosystems and food webs with species that are often already impacted by other anthropogenic stressors. We currently have very little idea what all of these changes to marine ecosystems mean for particular species except that many of those species are, in fact, responding to these changes.[164] As a result, we now live in a world of deep uncertainty about the current and future status of not only individual marine species but also entire marine ecosystems.[165] Within that uncertainty, however, it is clear that anthropogenic stressors are significantly reducing the ocean's resilience, primarily as a result of long-term failures in governance.[166] Moreover, at least some marine ecosystems, especially coral reefs and the Arctic Ocean, are almost certainly already transforming.[167]

Given these realities and the acknowledged failures of fisheries governance, pursuing MSY-based fishing goals calculated on the basis of old realities can only be viewed as a recipe for increasing the odds that all marine ecosystems will transform drastically and in ways that reduce the biodiversity and complexity of the ocean—as well as global human commerce, cultural integrity, and food security. As the Organisation for Economic Co-operation and Development (OECD) noted in 2010, "deterministic fisheries models [like MSY] . . . may have led some people to believe that sustainability of fisheries revolves around maintaining steady stock levels and steady catches over time. This is unlikely to be desirable for stocks the growth and reproduction of which depend critically on a fluctuating environment, and it may even be impossible to attain."[168] The MSY fisheries management framework embodies the traditional narratives of American natural resources law and policy, especially tragedy and sustainability, that do not fit the Anthropocene, and we need to reconfigure fisheries law—both domestically and globally—to deal with our new reality.

Incorporating Resilience Theory into Marine Fisheries Law and Policy

Resilience theory offers a different framework and approach for marine fisheries governance in the Anthropocene. First, as noted in Chapter 3, resilience theory counsels that change in natural systems is always expected as a result of complexly interacting adaptive cycles. Second, transformation of ecosystems is possible—and, as we have seen, already occurring in many marine environments. Third, as a result, management measures that work today may not work tomorrow, particularly if managers already know that disturbances are at work at multiple scales. For marine species, these disturbances include fishing, habitat destruction, pollution, climate change, and ocean acidification.

Most basically, a resilience-based management regime asks a fundamentally different policy question than our current fisheries laws. An MSY-based governance regime asks always, "How many fish can humans take without completely screwing up the ocean?" A resilience-based narrative asks instead, "In light of the current complex dynamics of the ocean in the Anthropocene and transforming marine ecosystems, how intensely—if at all—should humans be fishing the ocean, especially in light of other human dependencies on functional marine ecosystems and sufficient marine biodiversity?" In other words, the sustainability perspective assumes that continued exploitation of fish is possible without sacrifice of other marine values, while the resilience-based approach reassesses the human relationship to the ocean in a more holistic way, acknowledging that there may be trade-offs among our various needs and desires with respect to the ocean (oxygen, food, aesthetics, recreation, culture, religion, and so forth).

Putting these lessons into practice, we must begin by defining what we are trying to achieve in marine fisheries management in a climate change era—i.e., define the normative goals. As Chapter 3 noted, resilience theory offers little guidance here; ecological resilience and panarchy are system properties, not normative aspirations. Therefore, the political process behind law and policy could—even if fully informed by resilience theory—self-consciously choose to exploit all living marine resources as fast as possible before climate change-driven threshold crossings render the species that we currently depend upon commercially extinct.

The political force of short-term thinking is notoriously strong in the United States, and we are in danger of actually making this choice with respect to marine living resources, either as a conscious result of the "It's the End of the World as We Know It" narrative or, more insidiously, the unconscious pursuit of business as usual. However, any decision to continue to exploit living marine resources at current or even increased levels amounts to a disempowering relinquishment of responsibility. Such a choice essentially gives up on long-term marine management and *guarantees* the worst of all possible ocean futures when the ocean's fate is still deeply uncertain, throwing any pretense of intergenerational equity out the window in the process.

Let's posit instead that the most important goal of marine fisheries management in the Anthropocene should be to maintain functional marine ecosystems that are ecologically resilient to ocean acidification, climate change, and their complex synergistic interactions with other stressors, with a goal of promoting marine biodiversity over the long term. As part of this goal, ocean governance must acknowledge that marine ecosystems will change over time and that many of them will transform into new ecological states of being. Thus, when climate change and its synergies make ecological regime shifts inevitable, governance regimes should help to ensure, to the extent possible, that marine ecosystems and the SESs of which they are a part will transition to new states of being that are as productive and adaptive as possible, rather than collapsing into the equivalent of decimated marine deserts, jellyfish seas.

The studies predicting widespread fisheries collapses discussed above strongly suggest that this normative goal must prioritize ecosystem function over human exploitation. Continued fishing at current rates—both within the United States and globally—appears increasingly unsustainable under changing conditions. Instead, given this normative goal, the stressors humans have already imposed on the marine environment, and the deep uncertainties regarding the future of marine ecosystems, resilience theory counsels us to create governance systems that *minimize* human disturbances to these systems.

Unfortunately, many of the existing anthropogenic ocean stressors are not amenable to immediate reduction. Because of long lag times in the planetary climate and carbon systems, climate change and ocean acidification cannot be eliminated for centuries or a millennium, respectively. Existing coastal habitat destruction is unlikely to be reversed and will be complicated by rising sea-levels. New pollution of the ocean has been reduced, but removing legacy pollution in the form of the plastics, toxics, and nuclear waste poses a considerable technological as well as policy challenge. In contrast, reducing land-based pollution, especially contaminated runoff, is technologically feasible and can result in some fairly immediate improvements. The problem more often is that the political will and money to implement such measures are lacking. In the United States specifically, serious reductions of land-based pollution would require fairly stringent regulation of agriculture and power plants, both of which are industries that can mount (and repeatedly *have* mounted) considerable political and legal resistance to any such proposals.

Commercial fishing, on the other hand, is subject to immediate reduction, and reducing the allowable catch of particular species is a recognized strategy for improving their health.[169] Several organizations, scientists, and scholars have proposed that fisheries management needs to be far more precautionary, ecosystem-based, and flexible. All these experts recommend reduced catch limits.[170] Moreover, the goals of fisheries management must become more nuanced to reflect biological realities: "sustaining the resilience of fish populations requires that we seek to preserve their age and geographic structure rather than manage only their biomass."[171]

The United States has made some progress in implementing a more precautionary approach to fisheries management. The 2006 amendments to the Magnuson-Stevens Act imposed explicit duties on NOAA Fisheries and the regional FMCs to prevent overfishing and to rebuild overfished stocks.[172] Among other things, the amendments required Annual Catch Limits (ACLs) for federally managed fisheries by 2007.[173] In addition, these amendments have led to other kinds of fish conservation measures that fishers have repeatedly considered too draconian, including temporary shutdowns of certain fisheries, leading to several legal challenges. However, the federal courts have largely upheld these measures. For example, in *North Carolina Fisheries Association v. Gutierrez*,[174] the US District Court for the District of Columbia applied the new amendments to uphold new significant restrictions on the harvest of snowy grouper, vermillion snapper, and black sea bass despite uncertainties in the science, holding that "the Secretary [of Commerce] was not obliged to 'sit idly by' when faced with overfishing and overfished stocks simply because the data available to him may have been less than perfect."[175] However, while the sustainability index for US fisheries has generally been increasing since 2000, it is still worth noting that both the number of federal managed fish stocks subject to overfishing and the number of overfished stocks increased from 2014 to 2015.[176] Moreover, federal fisheries management remains grounded in MSY, posing increasing disjunctions between legal requirements and ecological realities for the future. Thus, while fisheries management since 2006 reflects some social and legal transformation in Americans' perspective on fisheries and their relationship to marine species, this social transformation is far from complete. Moreover, while many Americans are willing to support "sustainable fisheries," far, far fewer are willing to consider significant limitations on human exploitation of fish.

As a second resilience-based governance strategy, fisheries law should increase the use of marine protected areas, marine reserves, and marine spatial planning to protect areas of special biological importance and to reduce fishing pressures.[177] While reducing commercial fishing is always political, nations have been willing to do it to protect marine ecosystems that they acknowledge to be fragile. For example, many coral reef marine protected areas around the world either eliminate fishing entirely or severely regulate it. In the United States, both states (out to three miles from shore) and the federal government can establish marine protected areas through a variety of existing legal authorities.

Recent developments in the Arctic suggest that nations will also be willing to limit or eliminate commercial fishing in areas vulnerable to over-exploitation because of climate change. As Alaska fisheries demonstrate, the Arctic region offers rich fishing resources. However, the Arctic Ocean itself has not yet been subject to regular commercial fishing efforts because of the accessibility problems and other dangers that Arctic Ocean sea ice causes, even in summer. That fact is changing, however, because the Arctic is one of the regions of the world that is altering most because of climate change. As the National Snow and Ice Data Center reports, "the Arctic region is warmer than it used to be and it continues to get warmer. Over the past 30 years, it has warmed more than any other region on earth."[178] As part of this warming, Arctic Ocean sea ice has been disappearing. "Satellite data show that over the past 30 years, Arctic sea ice cover has declined by 30 percent in September, the month that marks the end of the summer melt season."[179] Scientists now predict that the Arctic Ocean could be ice-free in September as early as 2020, with the period of open ocean progressively extending.[180] The ice disappearance is occurring fast enough that the National Academy of Sciences considers it an "abrupt" change resulting from climate change.[181]

Because of the disappearing ice, the Arctic Ocean represents a future new place for commercial fishing, raising the specter of a mad rush of unregulated and destructive exploitation. Using the Magnuson-Stevens Act, in August 2009 the US Secretary of Commerce approved a far-sighted Fishery Management Plan for all parts of the Arctic Ocean under federal control, primarily the Chukchi and Beaufort Seas off of Alaska.[182] This plan "prohibits commercial fishing in the Arctic waters of the region until more information is available to support sustainable fisheries management."[183] The United States then led an international effort to precautionarily and peremptorily prohibit commercial fishing in all waters of the Arctic Ocean. In July 2015, the five nations that surround the Arctic Ocean—the United

States, Canada, the Russian Federation, Norway, and Denmark (via its holding of Greenland)—signed a declaration to prevent unregulated commercial fishing in the high seas (central) portion of the Arctic Ocean.[184] The declaration memorializes these nations' intention "to authorize their vessels to conduct any future commercial fishing in this area only once one or more international mechanisms are in place to manage any such fishing in accordance with recognized international standards. They also intend to establish a joint program of scientific research with the aim of improving understanding of the ecosystems of this area."[185]

More generally, other areas of the ocean are probably becoming particularly important for fostering marine ecosystem resilience to climate change and ocean acidification. Thus, in addition to increasing protections for areas such as nurseries and breeding grounds, nations should be looking to protect areas newly made important because of climate change impacts. For example, ocean warming tends to drive marine species toward the cooler waters around the poles. However, ocean acidification occurs most intensely in colder waters. "Sweet spots" are therefore likely to emerge in the more temperate regions of the ocean, where the effects of both ocean acidification and warming are attenuated enough to support diverse assemblages of species. These areas should be identified and protected.

Finally, resilience theory can prompt even more radical changes in marine fisheries management by shifting how we frame management choices. As noted, normal politics tends to evaluate changes in management only in the short term, framing efforts at long-term conservation as a choice between business as usual and forced economic damage to the fishing industry. A resilience-based perspective instead gives more weight to the longer-term view of fisheries management, refiguring the business-as-usual path as a choice that is increasingly likely to drive a growing number of marine ecosystems across ecological thresholds into unpredictable but probably less productive transformed states. From this new perspective, we can recognize that commercial marine fisheries are likely to decline in the next few decades regardless of what we do in law but that certain legal choices now can increase the odds that we will still have productive marine ecosystems at the end of the twenty-first century and beyond. As such, the most effective governance change to promote marine ecological resilience at this point is arguably to institute a worldwide phaseout of commercial wild marine fisheries (and possibly other significant kinds of fisheries, such as large recreational fisheries), coupled with an increased reliance on the well-regulated use of the more environmentally benign forms of marine aquaculture—a global industry that already has been on the rise in response to increasing demands for fish.

The marine trickster is a particularly powerful, unpredictable, and potentially disruptive trickster operating panarchically throughout the ocean. He will increasingly demand increased accommodation by human governance, including acknowledgement that we may not be able to pursue all marine SES priorities simultaneously in the same ways that we always have. Acknowledging a world of continual change and the significant lack of human control over complex ocean systems, resilience theory suggests a wide range of potential amendments to marine fisheries management for a changing ocean. Even the most modest of these, however, should inspire comprehensive amendments to both domestic and international fisheries law, particularly to their emphases on MSY. Full incorporation of resilience theory, in turn, demands a longer-term and system-based perspective on marine management. The adaptive capacity necessary to cope with this trickster, for example, should include increasing SESs' capacities to shift food sources.

This suggested shift in perspective is, indeed, demanding and likely to be resisted on several fronts. However, by empowering humans to make choices now to strengthen the ecological resilience of marine ecosystems to the changes that are still coming, we are empowering ourselves to increase the chances that the ocean will remain a complex and biodiverse natural system far into the future.

CHAPTER SIX

Thinking Like a System

Resilience as a Narrative of Connection

Up to this point, we have introduced the concept of ecological resilience and the Holling school of resilience theory as a new environmental paradigm for the Anthropocene. We contrasted resilience with previous narratives that humans have used to describe their relationship to the environment and provided some examples of the benefits of a resilience narrative through case studies. In this chapter, we examine an underlying—and to date unrealized—potential for resilience theory to transform our relationship to the Anthropocene.

We acknowledge that most applications of "resilience" approaches so far look much like previous approaches to natural resource management in the sense that they take a control-based approach to governance. Here, we suggest that a true embrace of social-ecological systems (SESs) resilience theory can and *should* be much more radical. Resilience has the potential to reconfigure how we think about "social-ecological" systems by breaking down this binary and embracing the reality that we *are* these systems. There are no social elements that can be separated from their ecological context, just as there are no longer places in nature that have not been influenced by decisions made by human societies. As Aldo Leopold noted in 1935,

One of the anomalies of modern ecology is that it is the creation of two groups, each of which seems barely aware of the existence of the other. The one studies the human community almost as if it were a separate entity, and calls its findings sociology, economics, and history. The other studies the plant and animal community, [and] comfortably relegates the hodge-podge of politics to "the liberal arts." The inevitable fusion of these two lines of thought will, perhaps, constitute the outstanding advance of the present century.[1]

In the title of this chapter, we invoke Aldo Leopold's iconic concept of "thinking like a mountain" in order to call for the inclusion of communitarianism as a critical component of resilience theory, adaptive governance, and resilience thinking moving forward. Leopold was a resilience scholar before the concept even existed. His key message was that humans are a part of a land community. He argued that the only way to meaningfully address environmental challenges is to develop what he called a "land ethic." As Leopold explained, "The land ethic simply enlarges the boundaries of the community to include soils, waters, plants, and animals, or collectively: the land."[2] This idea was radical for his time and is still radical today. By adopting some of the core principles of Leopold's land ethic and other community-based perspectives on human relations with the environment, resilience theory can realize its untapped potential.

We acknowledge that this endeavor requires resilience scholars and practitioners to engage the normative implications of their work. As we have noted in previous chapters, resilience theory is not normative in the sense that the theoretical framework itself offers no answer to whether maintaining a given system is good or bad. Nevertheless, the moment the theory is placed into some governance context involving the real world, values and beliefs about how the world should be inevitably come into play. As Chapter 5 emphasized, for example, it is not enough to say that we will make resilience a goal of marine fisheries management; we must far more precisely state whether we are prioritizing the long-term resilience of ocean ecosystems in the face of climate change or the short- and medium-term resilience of commercial fisheries to changing marine ecologies. Those are *not* the same governance goals, and they result in different governance strategies. Failure to address the trade-offs of (and between) strategies and the critical normative decisions for governance validates one of the prominent critiques of resilience theory—that it often fails to acknowledge the complex societal dynamics involved in natural resources management. This critique is often summarized as a failure to ask the question: resilience for whom? We agree with these critics that any meaningful governance application of resilience theory *must* ask this question.

Our argument, however, goes one step further. In a governance context, answers to the core normative questions of resilience theory are almost always anthropocentric—focused on the needs of the humans that use system resources. If you were shocked at the suggestion in Chapter 5 that resilience theory counsels us to consider phasing out all commercial and other large-scale marine fisheries worldwide, then you have felt the pervasive grip of this anthropocentric perspective on natural resources management. We agree with Leopold, who argued that such anthropocentric thinking is at the core of our environmental troubles.

Communitarianism, in contrast, seeks to rebalance the liberty and autonomy of individuals with the values of functional, supportive, and productive communities. Communitarian scholars thus generally present their work as a profound critique of radical liberal individualism, reminding us that individuals can thrive only within broader community structures and networks. Political philosopher Daniel Bell argues that:

> Neither human existence nor individual liberty can be sustained for long outside the interdependent and overlapping communities to which we belong. Nor can any community long survive unless its members dedicate some of their attention, energy, and resources to shared projects. The exclusive pursuit of private interest erodes the network of social environments on which we all depend, and is destructive to our shared social experiment in democratic self-government.[3]

Like the scholars who call themselves "new communitarians," we are concerned "with the balance between social forces and the person, between community and autonomy, between the common good and liberty, between individual rights and social responsibilities."[4]

We also agree with communitarian philosophers that the definition of "community" is critical. Bell acknowledges that the least developed part of communitarian philosophy in its debate with liberalism is "questions about which communities we are to value and the political implications that flow therefrom."[5] Amitai Etzioni describes "community" as a multi-scalar system of "webs of social relations that encompass shared meanings and above all shared values," ranging from families to villages and neighborhoods to nations to even international ethnic subgroups, invoking a panarchy-like vision of interacting sources of meaning and value, which both influence and are influenced by the individuals who are embedded within these webs.[6] Like Leopold, we argue that the community we should value and protect includes the myriad of ecological systems that US society tends to view as nonhuman and separate but on which we are utterly dependent. Like our social and cultural communities, we both influence and are influenced by these systems, and both our individual identities and social structures depend upon our continued and supportive networking with our ecological communities.

While a community-based perspective may seem to be at odds with the United States' cultural and legal focus on individualism and individual rights, including property rights, communitarian principles have always been an important component of US law and policy—particularly in times of change and stress. Legal doctrines such as the public trust, nuisance, and necessity privilege the rights of the community over the desires of individuals, acknowledging that public health, collective well-being, and the survival of the community take precedence over individual liberty and autonomy. State public trust doctrines, for example, subordinate private interests in navigable waters to the public's rights to fishing, navigation, commerce, and (usually) recreation, preventing private riparian landowners from blocking or even charging for these uses. Public nuisance limits the uses that private landowners may make of their properties to protect the larger community, especially in terms of how much noise, pollution, and danger individual property owners may create. The doctrine of necessity is a broadly applicable communitarian doctrine. For example, it authorizes both mandatory vaccination and, even more stringently, quarantine measures during outbreaks of contagious and deadly diseases. Quarantine by definition deprives individuals of liberty, but it reflects society's judgment that the rights of infected individuals must be limited by the rights of healthy individuals and especially the needs of society as a whole.

Quarantine, of course, is a particularly stark example of how the law effectuates communitarian values at the expense of individual rights in extreme survival contexts. To illustrate that legal communitarian balancing can be far more nuanced, this chapter will examine one of the most contentious areas of law in the United States: The regulatory "takings" doctrine. The Fifth and Fourteenth Amendments to the US Constitution, and similar provisions in state constitutions, forbid governments from taking private property for public purposes without paying just compensation. In a classic takings or condemnation case, the government intends to physically use private real property for a highway, public park, or government building, and it is well settled in law that these are valid public purposes for which the government can indeed condemn private property, so long as it pays just compensation. Beginning in 1922, however, the US Supreme Court began to recognize so-called "regulatory takings," allowing that the mere regulation of property could also require compensation if it goes "too far."[7] Environmental and natural resource regulation have been frequent targets of regulatory takings claims.

However, looking more closely at the interface between environmental protections and property rights reveals that US law is already more communitarian than most Americans realize. By examining the case law that analyzes when environmental and other protections become a "taking" of property under the Fifth and Fourteenth Amendments, we reveal the underlying challenges that arise through the emphasis on a "rights" framework to describe our relationship to land, demonstrating that the relationships among individuals, their communities, and the environment of which they are simultaneously both a part and depend upon are far more complex, nuanced, and system-based than a purely rights-based narrative can adequately describe or patrol. After a review of Leopold's land ethic and its potential role in resilience thinking, we turn to case law regarding Fifth Amendment takings to demonstrate the tensions between rights and relations and the need for a broader, more communitarian approach moving forward.

Aldo Leopold: The First Resilience Scholar?

Aldo Leopold was a groundbreaking American thinker and conservationist. He is best known for his book *A Sand County Almanac,* in which he puts forward the land ethic, among other essays. Born in 1887 and raised in Burlington, Iowa, Leopold developed an interest in the natural world at an early age, using his mother's opera glasses to identify birds and other species. He graduated from the Yale Forest School in 1909, then took a position with the newly established US Forest Service in Arizona and New Mexico. He was the first supervisor (at the age of twenty-four) for the Carson National Forest in New Mexico. In that position, he successfully proposed the creation of the first wilderness area, which was established in the Gila National Forest in 1924. Following a transfer to Madison, to the US Forest Service's Forest Products Laboratory, in 1933 he became the first professor of game management at the University of Wisconsin. During that time, he published the first textbook on wildlife management and, in 1935, he bought the Sand County farm that provided the setting and many observations for the ecological essays published in *A Sand County Almanac* a year after his untimely death in 1948. For all these reasons and more, Leopold is considered to be the father of many groundbreaking ideas: wilderness, wildlife management, ecology, and conservation.

Leopold saw firsthand what is perhaps the most devastating legacy of the Manifest Destiny narrative: the Dust Bowl. The Dust Bowl era, a decade-long series of severe dust storms in the 1930s in the American midwestern prairielands, was one of the largest ecological disasters in human history. Its root causes were a combination of Euro-American farming practices and environmental bad luck. The plowing under of prairie to grow wheat and other products removed the deep-rooted grasses that normally kept the soil in place. At the same time, a severe drought in the region stressed the system and contributed to the loss of topsoil. In his book on the Dust Bowl, Donald Worster explains that, without the natural prairie grass to keep the soil in place, the land dried, turned to dust, and blew away eastward and southward in large dark clouds. At times the clouds blackened the sky, reaching all the way to places such as New York and Washington, DC. Much of the soil ended up deposited in the Atlantic Ocean, carried by prevailing winds.[8] Recurrent dust storms wreaked havoc, choking cattle and pasturelands and driving 60 percent of the population from the region. Occurring around the same time as the Great Depression, it was a bleak and difficult time in American history. Much of Leopold's career was spent trying to convince foresters and farmers to employ soil conservation measures to avert such crises—or, as Leopold more eloquently put it, to take on "the oldest task in human history: to live on a piece of land without spoiling it."[9]

Leopold's experiences in wildlife and land management over the course of decades, including his own farm in Sand County, convinced him that the solution to soil erosion and other environmental management challenges did not reside in environmental education, technological advancement, or government action. Something more fundamental was required. Leopold's land ethic called upon humans to think of themselves as a part of a larger biotic community that includes not only animals but also soils, water, and plants. This community, referred to collectively by Leopold as "the land," is what has been referred to in this book and throughout resilience theory as a complex social-ecological system (SES).[10] However, Leopold's invocation of community does something that resilience theory currently does not: It collapses the social/ecological binary. It also calls upon humans to engage in an ethical relationship with other elements of the land system. "That land is a community is the basic concept of ecology, but that land is to be loved and respected is an extension of ethics."[11] The ethical obligation is not restoration but overall health: "A land ethic, then, reflects the existence of an ecological conscience, and this in turn reflects a conviction of individual responsibility for the health of land."[12]

The ethical call to "think like a mountain" is to embrace the complexity and interconnectedness that makes up our relationship to the land. The phrase reflects Leopold's own transition from a *manager* to a *member* of the land community. In *A Sand County Almanac*, he reflected on his own role in removing wolves and other predators from wild places in New Mexico and elsewhere to accommodate cattle grazing and increase deer herds for human hunters. The ecological unbalance that occurred caused him to reflect on the hubris associated with focusing on parts of the system rather than its whole:

> I have lived to see state after state extirpate its wolves. I have watched the face of many a newly wolfless mountain, and seen the south-facing slopes wrinkle with a maze of new deer trails. I have seen every edible bush and seedling browsed, first to anaemic desuetude, and then to death. I have seen every edible tree defoliated to the height of a saddle-horn. . . . In the end the starved bones of the hoped-for deer herd, dead of its own too-much, bleach with the bones of the dead sage, or molder under the high-lined junipers. . . . So also with cows. The cowman who cleans his range of wolves does not realize that he is taking over the wolf's job of trimming the herd to fit the range. He has not learned to think like a mountain. Hence we have dustbowls, and rivers washing the future into the sea.[13]

Leopold invoked both predator removal and the Dust Bowl as examples of the dangers associated with not understanding the land as a whole.

In "Leopold's Last Talk," Eric Freyfogle reports on his archival research into over 150 public lectures Leopold gave in his later years, distilling them into the essence of Leopold's message. Leopold's message is, in many ways, much more radical than it appears in *A Sand County Almanac*.[14] Professor Freyfogle points out how nonconformist the land ethic is, placing Leopold within his historical context and outlining the "centuries-long intellectual trajectory" of humanism and liberalism:

The then-ascending impulse was for humans to rise above nature, seeing it as a complex but ultimately knowable machine and controlling it in service of human wants. It was an impulse—grounded on the humanist side of the Renaissance—that gave rise in complex ways not only to advances in science and technology, but to the revolutions of the seventeenth and eighteenth century, the emergence of economic liberalism, and the expanding embrace of individual rights. Put simply, the independent thinker of the age of Descartes (early seventeenth century) had matured into the morally autonomous, utility-seeking actor of the age of Bentham and J. S. Mill (nineteenth century), and gone onward to become the rights-bearing, vote-wielding citizen of the early twentieth-century. In the emergent liberal ideal, an individual could act as she saw fit, crafting and pursuing a self-chosen vision of the good life so long as she caused no material harm and recognized the equal rights of others to act similarly. Nature was where human life unfolded, and science helped guide its manipulation.[15]

Leopold's land ethic departs radically from the humanist, liberal paradigm. According to Freyfogle, Leopold "called for profound changes in not just the liberal traditions of individual autonomy and economic liberty but the main components and dualities of Enlightenment thought."[16] Leopold valued scientific knowledge but also recognized its limits, calling for an ethical orientation instead of a scientific one.

Freyfogle identifies several themes within Leopold's message that resonate with resilience theory. For example, Leopold's frequent use of "health" as an environmental goal for the land community can be seen as another way of assessing system dynamics. Invoking health, he used the human body as an analogy. In one lecture, he stated: "I must ask you to think of land and everything on it (soil, water, forests, birds, mammals, wildflowers, crops, livestock, farmers) not as separate things, but as parts—organs—of a body. That body I call the land (or if we want a fancy term, the biota)."[17] In another, he referred to land as the most complex of all organisms.[18] In many ways, what Leopold referred to as land health would be described by a resilience scholar as self-organization and emergent properties, two key aspects of the resilience of complex systems. "Basic to all conservation is the concept of land-health; the sustained self-renewal of the community."[19]

Leopold's thinking also differs from the majority of current resilience scholarship in several important ways. In Leopold's complex system, there is no categorization dividing the human and the nonhuman. Instead, the community includes us all. Resilience scholars usually describe human/ social systems (people, governance, built infrastructure, culture) as connected with ecology (climate, biodiversity, forests, ocean), but there is still a reinforcement of the idea that the "social" and the "ecological" are fundamentally different (if overlapping) systems. By collapsing this binary, Leopold's thinking is both more radical and more accurate.

This collapse relates closely to the second key difference between Leopold's thinking and that of most resilience scholarship: the rejection of anthropocentrism. While most resilience scholarship is not explicitly anthropocentric, the focus is almost always on how social and ecological processes can better support human life and the needs of humans over time. Leopold saw humans as citizens of the land community, responsible for our role in the larger community but also inextricably bound to it.

Leopold was also very skeptical about the use of economic approaches to solve environmental problems, emphasizing the need to quit thinking about decent land use as solely an economic problem.[20] While recognizing the influence that economics has on decisions, Leopold understood that our economic well-being could not be separated from the well-being of our environment. At the time of his writing, the nation was very much swept up in the Manifest Destiny narrative, which combines human exceptionalism with economic prosperity in order to make "progress" inevitable. Worster's three-part formula for the mind-set that contributed to the Dust Bowl era underscores Leopold's point: (1) Nature is seen as capital; (2) humanity has a right to use capital for self-advancement; and (3) the social order should permit and encourage this continual increase of personal wealth.[21] Having seen the degradation of private lands created by this narrative first hand, Leopold remained skeptical of free-market solutions. He saw the concept of "property rights" as a license for owners to ignore the larger implications of their activities. As he explained in *A Sand County Almanac*, "We abuse land because we see it as a commodity belonging to us. When we see land as a community to which we belong, we may begin to use it with love and respect."[22] Freyfogle notes that Leopold never made normative use of individual-rights rhetoric and that he viewed property rights in particular as subordinate to the common good of land health. In broadening the idea of "community" to include sentient and nonsentient beings, Leopold did not employ the concept of rights, as many do today.[23]

Leopold's rejection of individual, rights-based liberalism came from his understanding that the Manifest Destiny narrative needed to be rejected at its core. Instead of seeing nature as capital, it should be seen in all its diversity as a coinhabitant of the earth, a fellow citizen of the planet. Rather than seeing humans as having the right to use capital (nature) for self-advancement, humans should assume their ethical duty to strive to maintain land health. Finally, instead of the social order permitting and encouraging the continual increase of personal wealth, society should facilitate personal connections with that land.

Leopold's call for a cultural shift from land-as-commodity to land-as-community seems as relevant—and as radical—today as it was when *A Sand County Almanac* was published in 1949.[24] If anything, the privatization approach to environmental protection and the commodification of nature have only increased in their influence. Particularly in the context of the developing world, neoliberal approaches by the World Bank and other institutions are dominant, commodifying everything from water, fish, and minerals to carbon sequestration.[25]

It is along these lines, among others, that some critics of resilience theory have concerns. At this point in the development of the literature, resilience theory and practice both tend to normalize many environmental approaches that commodify nature and to ignore the complexity of contemporary and historical social processes involved in SES dynamics.[26] In contrast, a Leopoldian communitarian approach to resilience theory, especially in the Anthropocene, encourages natural resources governance institutions to think in terms of promoting the health of the entire system—whether that means adapting to a transforming forest ecosystem in New Mexico's Rio Grande Basin or considering the possibility that the best thing for the health of the ocean system is to transition away from large-scale commercial fishing.

What Would Leopold Say about Resilience Theory?

Leopold's land ethic has much in common with modern resilience theory. Both embrace the complexity of systems. Both assess the health or self-renewal capacity of systems and place an emphasis on understanding system function and dynamics. As one of the first ecologists, Leopold understood the importance of what he called "intelligent tinkering"—management of the land that included valuing and keeping all its parts. Resilience theory shares this view, with the caveat that, sometimes, parts are lost whether we want to keep them or not. When transformation occurs, the community must embrace a new identity. Similarly, the emphases in resilience theory on complexity and continual learning are reflective of Leopold's thinking. He often remarked upon the need to understand the limits of science: "The ordinary citizen today assumes that science knows what makes the community clock tick; the scientist is equally sure he does not. He knows that the biotic mechanism is so complex that its workings may never be fully understood."[27]

What would Leopold make of the Anthropocene? Global climate change presents a challenge for the land community beyond anything Leopold encountered during his time. Leopold's call for a land ethic reflects in part his own struggles with the inaction he experienced during the Dust Bowl, to deforestation and other environmental challenges of his day. He states:

> One of the penalties of an ecological education is that one lives alone in a world of wounds. Much of the damage inflicted on land is quite invisible to laymen. An ecologist must either harden his shell and make believe that the consequences of science are none of his business, or he must be the doctor who sees the marks of death in a community that believes itself well and does not want to be told otherwise.

Denial of environmental challenges is not new. However, it combines with the Manifest Destiny narrative to create in humans a sense of separation from the natural world.

While Leopold would no doubt appreciate much of what resilience scholars have been able to accomplish, it is likely that he would also point out some of the limitations, or even flaws, in the development of resilience theory to date. The first point is perhaps the most obvious. Resilience theory conceptualizes a dynamic social-ecological system, whereas Leopold collapses this duality and calls upon humans to take their place within the larger community. Leopold's vision requires extending our ethical concern beyond the human to other community members, including rivers and streams, other sentient creatures, soils, and so forth.

Given the recognition by resilience scholars that SESs are inextricably linked, it would be tempting to discard Leopold's call for community, arguing that the distinction is mere semantics. However, the view that humans are separate from nature runs deep in our culture. Associated with it is an assumption that the more humans are involved, the less "natural" something is. This assumption is why wilderness is considered more "natural" than farmland, farmland more "natural" than a parking lot, and so forth. Leopold's land ethic calls for a paradigmatic change in our perception. Humans *are* nature. Likewise, nature *is* social in the sense that our collective agreements about what is natural create the categories of wilderness, parkland, farmland, and parking lot. There is no place where one exists without the other. Leopold brings humans into the land community in order to accomplish an extension of ethics to the nonhuman.

This is the second basis for Leopold's likely critique of resilience theory. As discussed, resilience theory is not normative in the sense that the theoretical framework itself holds no position regarding whether maintaining a particular system or system state is good or bad. However, resilience theorists must recognize that the moment the theory is placed in some context, values come into play. Several critics argue that resilience scholars often ignore the complexity of the societal and ethical implications of their characterizations of SESs.[28] The political and economic structures and the current and historical allocations of power or wealth within a society are critical to understanding SESs. When the complexities of these issues are not addressed, characterizations of SESs tend to reflect the humanist, Enlightenment paradigm that dominates Western culture today every bit as much as during Leopold's time.

Take, for example, the UN's Millennium Ecosystem Assessment (MEA).[29] The MEA of 2005 was a monumental effort to gather all relevant information regarding the state of the planet and was the result of thousands of experts from ninety-five countries assessing how ecological systems and processes influence human well-being. In its general report, *Ecosystems and Human Well-being: General Synthesis,* the MEA team framed its findings in terms of ecosystem services, defined as:

> the benefits people obtain from ecosystems. These include provisioning services such as food, water, timber, and fiber; regulating services that affect climate, floods, disease, wastes, and water quality; cultural services that provide recreational, aesthetic, and spiritual benefits; and supporting services such as soil formation, photosynthesis, and nutrient cycling.[30]

Conclusions were based on scientific findings that noted varying degrees of certainty and were tied to the UN's Millennium Development Goals, which in turn are all tied to improvements in human well-being, such as reducing child mortality, promoting education, and eradicating extreme poverty—with the one exception being the goal to "ensure environmental sustainability."[31]

Of course, all of these goals are very important. The point is not to set aside human needs. However, when we focus on ecological systems only in terms of how they serve us, we greatly reduce our understanding of the complex world in which we live, and we are more likely to undervalue the various system elements and dynamics that support SES resilience. As Leopold noted:

> One basic weakness in a conservation system based wholly on economic motives is that most members of the land community have no economic value. Wildflowers and songbirds are examples. Of the 22,000 higher plants and animals native to Wisconsin, it is doubtful whether more than 5 percent can be sold, fed, eaten or otherwise put to economic use. When one of these non-economic categories is threatened, and if we happen to love it, we invent subterfuges to give it economic importance.[32]

While the concept of market-based, service-oriented environmentalism is more sophisticated now than it was in Leopold's day, his basic premise still holds true. We cannot begin to grasp the complexity of the various interactions among the land community and the role they play.

In the US context, there are few assessments of ecological processes required by law that are not directly related to human needs. Most assessments at the federal level take place under the National Environmental Policy Act (NEPA), which is triggered only when there is a proposed federal action that might have a "significant impact on the human environment."[33] These assessments, called "Environmental Impact Statements," do not require protection of the environment.[34] They simply facilitate development with the hope of creating a more informed and (ideally) less environmentally damaging project. One notable exception to this anthropocentric view is the Endangered Species Act, which values species themselves. It is arguably the most controversial environmental law in the United States, and its refusal to look at human needs when deciding whether to list a species is a main reason.[35] Otherwise, attempts in the United States to address biodiversity issues more generally have failed. The United States remains one of four nations *not* to ratify the United Nations Convention on Biodiversity, for example. Moreover, in 1993, the Clinton administration proposed the National Biological Survey, a new federal agency tasked with taking an inventory of the ecological health of the nation.[36] The proposal was abandoned when fears regarding the possible outcomes from government investigations on private land made it untenable.

The Holling school of resilience theory is born of the humanist, Enlightenment paradigm, though many of its ideas undermine that paradigm's legitimacy. Its acknowledgment of the complexities and dynamics of systems vitiate overly simplistic notions of agency. Bruce Braun notes the radical potential of resilience and other complex systems theory to reconfigure our notion of the material world, yet he cautions that "by failing to reflect on the history of its own ideas [resilience theorists] risk failing to distinguish between the original critical impulses of complex systems theory and its notions of non-deterministic nature, and modes of neoliberal governance in which these ideas are absorbed and redeployed."[37]

The "modes of neoliberal governance" Braun refers to include payments for ecosystem services projects, the privatization of land and water, and other related management approaches that increasingly dominate environmental governance in the Anthropocene. Applications of resilience theory often (and without reflection) integrate both liberal and neoliberal notions of nature as "capital" and ecosystems as a source of "services" for humanity. These characterizations of the land community are so pervasive that, to some extent, to blame resilience scholars for integrating these ideas is akin to blaming the fish for the quality of the water in which it swims. However, it is important that resilience scholars become more cognizant of the ways in which these characterizations are dangerous because they perpetuate the commodity-driven culture that created the Anthropocene. Resilience scholars must acknowledge the inherently normative quality of applying resilience theory and make more conscious choices about what assumptions about nature and culture are perpetuated in practice.

If persuaded by Leopold that—at its core—environmental degradation is a matter of ethics, the employment of anthropocentric and neoliberal characterizations of SESs is inherently problematic. Fortunately, Braun and others have recognized that, to the extent these characterizations occur, they highlight an issue regarding the *application* of resilience theory rather than with the theory itself.

Importantly, therefore, while most current applications of resilience theory look much like previous approaches to natural resources management and reflect the humanist, Enlightenment paradigm, the resilience framework for SES dynamics has the capacity to capture the complexity existing within both social and ecological systems, including the value systems involved. It also has the capacity to de-center the "human" within the system and place our species and accompanying actions within the larger complexity of SESs. As such, resilience theory has the potential to reconceptualize how we think about "social-ecological" systems by breaking down this binary and embracing the reality that we are deeply and inextricably embedded in complex systems. Leopold was attracted to the work of John Dewey and other American pragmatists, a philosophy that places emphasis on lived experience rather than abstract ideals. The lived experience of the land community taught him that we are all connected. The land ethic is a narrative of connection and mutual influence. In reality, the lived experience of property has also always been relational, taking into account not only the liberal notion of "rights" but also collective societal goals.

Rights and Relations in Property Law

The concept of property rights and its evolution in the United States provides an illustration of the extent to which Leopold's land community is already reflected in our laws and policies. It also shows how the governance of property rights and natural resources use in the United States has always been more communitarian than generally perceived, providing a basis for rethinking the role of law in creating the land and water community. Both during Leopold's time and today, there is a general belief that private rights are grounded in the Constitution and that limitations on private land use are legitimate only when private actions visibly harm neighbors or the surrounding community. The reality is much more complex because we *are* a part of a community.

In its founding, the United States embraced Enlightenment ideals, including its rhetoric of rights. In some respects, ambitions of property ownership drove the creation of the new nation-state.[38] Property rights are reflected in the Fifth Amendment to the US Constitution, which requires the federal government to compensate landowners when property is "taken" for public use without compensation. After the Civil War, this federal constitutional requirement was extended to the states through the Fourteenth Amendment and the US Supreme Court's conclusion that that amendment incorporated Fifth Amendment protections for citizens against states.

The Takings Clause puts forth a seemingly straightforward idea. However, while the concept of requiring compensation is relatively simple, the legal determination regarding when a "taking" of property has actually occurred forms one of the most complex and dynamic fields of constitutional law. This legal complexity is a reflection, at least in part, of the basic fact that the lived experience of property ownership is more complicated than is generally perceived. The evolution of the case law regarding the Takings Clause reveals the ongoing tension between the liberal ideals embedded within the US Constitution and the practical reality of living as part of a community. After a brief overview of US Supreme Court interpretations of the Takings Clause to date, this section demonstrates the problems associated with employing the concept of "property rights" and the way in which *relationship* is actually a more appropriate framework for conceptualizing interactions among cultural expectations, property owners, and the land community.

As constitutional amendments go, the Fifth Amendment is actually a bit of a grab bag. It states:

> No person shall be held to answer for a capital, or otherwise infamous crime, unless on a presentment or indictment of a Grand Jury, except in cases arising in the land or naval forces, or in the Militia, when in actual service in time of War or public danger; nor shall any person be subject for the same offense to be twice put in jeopardy of life or limb; nor shall be compelled in any criminal case to be a witness against himself, nor be deprived of life, liberty, or property, without due process of law; *nor shall private property be taken for public use, without just compensation.*[39]

After all kinds of rights about criminal prosecutions, including grand jury investigation, the prohibition of double jeopardy, and the right to "due process," comes this amendment's Takings Clause: "nor shall private property be taken for public use, without just compensation."[40] Importantly, the Takings Clause does not prohibit the government from taking property when needed—it simply requires the government to pay market value in return. From the beginning, the Takings Clause recognized the need for public use to occasionally supersede individual interests for the benefit of the larger community, such as when there is a need to widen roads, add flood control infrastructure, or construct public buildings.

Over the decades, takings jurisprudence has evolved along with the nation to reflect the changing needs of society as well as the capacity of humans to alter the natural world. As discussed in Chapter 1, the Industrial Revolution brought with it air pollution, mining, and other environment-altering activities that created a need for government control over industrial practices. When the founders created the Takings Clause, however, the only type of "taking" envisioned was a physical taking—that is, the government's actual use and occupation of land (or, occasionally, other forms of property, such as when the federal government condemned patents during World War II to support the war effort).[41]

However, since 1922, the US Supreme Court has held that there are two types of takings requiring compensation under the Fifth and Fourteenth Amendments: physical and regulatory. Physical takings are relatively straightforward, the classic example being a road building project in which the government acquires titles to private property in order to physically construct an amenity that the public needs. In *Loretto v. Teleprompter Manhattan CATV Corporation* (1982), the Court determined that any form of physical invasion of private property by the government or at the government's command constitutes a "taking," even a relatively small invasion. The case involved a law by the City of New York requiring landlords to accommodate the installation of cables on their private property so that cable companies could provide services to tenants. Even though the physical invasion was a small (consisting of two cable boxes and some wire on a rooftop), a taking had occurred because of the "traditional rule that a permanent physical occupation of property is a taking."[42] *Loretto* now stands for what is called a per se rule—any permanent physical occupation of property is a constitutional taking. In some contexts, however, physical takings can be a bit blurry. In the water rights context, for example, lower courts have struggled over whether restrictions on water use to protect species constitute a physical or a regulatory taking.[43]

Regulatory takings are more difficult to determine, and they represent an incursion into the original balancing between public and private interests in property. Until 1922, any proper regulatory exercise of the police power—the government's general power to regulate to protect public health, safety, and welfare—could not be deemed a taking. Even after 1922 the US Supreme Court upheld as constitutional the new institutions of land use planning and zoning, recognizing that the Takings Clause's protections had to be balanced against the changing realities of modern society:

> Building zone laws are of modern origin. They began in this country about 25 years ago. Until recent years, urban life was comparatively simple; but, with the great increase and concentration of population, problems have developed, and constantly are developing, which require, and will continue to require, additional restrictions in respect of the use and occupation of private lands in urban communities. Regulations, the wisdom, necessity, and validity of which, as applied to existing conditions, are so apparent that they are now uniformly sustained, a century ago, or even half a century ago, probably would have been rejected as arbitrary and oppressive. Such regulations are sustained, under the complex conditions of our day, for reasons analogous to those which justify traffic regulations, which, before the advent of automobiles and rapid transit street railways, would have been condemned as fatally arbitrary and unreasonable. *And in this there is no inconsistency, for, while the meaning of constitutional guaranties never varies, the scope of their application must expand or contract to meet the new and different conditions which are constantly coming within the field of their operation. In a changing world it is impossible that it should be otherwise.*[44]

Given careful and well-documented studies that zoning in urban areas would protect public health and increase public safety—particularly the safety of children—the Court had no trouble concluding that comprehensive zoning was a legitimate exercise of cities' police powers, despite imposing new restrictions on property owners' uses of their properties.

Occasionally, however, land use regulation and environmental protection seemed to the Court to effectively target particular properties, making those property owners bear the burden of protecting the public at large. This is the perception of unfairness that lies at the heart of the regulatory takings doctrine—while it might be constitutional to limit all property owners a little to protect the public at large, it's not legitimate to significantly burden only a few property owners to achieve the same ends. The first regulatory takings decision in *Pennsylvania Coal Co. v. Mahon* reflects this sense of constitutional unfairness. Pennsylvania Coal Company owned the right to mine coal underneath Mahon's home, the act of which would have caused the house to sink as a result of subsidence. While the company had a property right to mine,[45] Mahon argued that Pennsylvania's Kohler Act regulated such practices and prohibited mining that caused such damage. The US Supreme Court held that the Kohler Act violated the incorporated Fourteenth Amendment Takings Clause, finding that Pennsylvania exceeded its police powers. Writing for an 8-1 majority, Justice Oliver Wendell Holmes, Jr., explained: "The general rule is that while property may be regulated to a certain extent, if regulation goes too far it will be recognized as a taking."[46]

So when does a regulation go "too far?" *Mahon* was a relatively easy case, because the Kohler Act completed destroyed Pennsylvania Coal's property right. Since then, however, the Supreme Court has emphasized that there is no "set formula" for determining when a regulatory taking has occurred, and it employs a difficult-to-predict balancing test to decide that question. Decades after *Mahon*, in *Penn Central Transportation Company v. City of New York* (1978), the Court provided some general guidance. The case involved a proposal by a railroad company to construct a high-rise building atop Grand Central Terminal. The City of New York had regulations in place to protect cultural landmarks, and its Landmark Preservation Commission rejected Penn Central's proposal.[47] The Court agreed with that decision, noting that "to balance a 55-story office tower above a flamboyant Beaux-Arts facade seems nothing more than an aesthetic joke."[48] It further held that the city's cultural protections were not a constitutional taking because the economic impact on Penn Central was not severe enough: Penn Central could still continue to use the station and the regulation did not interfere with the company's reasonable investment-backed expectations. The Court created a three-prong analysis for regulatory takings involving examination of: (1) the regulation's economic impact on the claimant; (2) the extent to which the regulation interferes with distinct investment-backed expectations; and (3) the character of the government action. In practice, application of the *Penn Central* factors rarely results in the finding of a taking, and the US Supreme Court has never found a taking using this formula.

Nevertheless, the Supreme Court still views regulations that destroy all of the value of private property with great suspicion. Its leading case finding a taking based on a regulation's economic impact is *Lucas v. South Carolina Coastal Council* (1992). *Lucas* involved a real estate developer who purchased two beachfront lots in South Carolina. After Lucas purchased the properties, South Carolina enacted the Beachfront Management Act, which barred Lucas from erecting any permanent habitable structures on his lots. Lucas claimed that a taking had occurred because of the government's restriction, and the defendants—damningly, in retrospect—conceded that the new statute destroyed all economic value of the property. Writing for the majority, Justice Scalia held that a regulation that results in total deprivation of a property's beneficial use is the equivalent of a physical appropriation and therefore a per se taking—the *Penn Central* balancing test does not apply. The Supreme Court further held that the state must provide compensation unless it can identify background nuisance and property rules that would prohibit the proposed activities. Importantly, however, the Court did not find that a taking had actually occurred but instead remanded the final decision to the lower court for a factual determination of whether the regulation had indeed deprived Lucas's property of all use or value. This never happened. After the Supreme Court's decision, South Carolina settled the case out of court. The *Lucas* case is significant for precedential purposes, however, and has come to be known for creating the second per se rule in the takings jurisprudence: when a regulation results in the loss "all use or value" of the property, it constitutes a constitutional taking and requires compensation.

Lucas is also an example from one set of regulatory cases that is particularly relevant for climate change and the Anthropocene—the coastal takings cases, of which the *Lucas* decision is the most extreme. Of course, as is true with regulatory takings claims in all contexts, most of regulatory takings claims based on coastal regulations fail, either for procedural issues[49] or on the merits.[50] On the merits, coastal regulatory takings claims can fail for a variety of reasons, including the state's application of background principles of state property law.[51] More interesting, however, is the growing recognition among the courts in coastal states that coastal properties are inherently vulnerable and that this vulnerability has bearing both on the regulatory takings analysis and the compensation owed even in a physical taking situation. One can see in these cases a nascent and only slowly emerging coastal land ethic—an evolving perspective on coastal land that is significant and worth encouraging. For example, in *Gove v. Zoning Board of Appeals of Chatham,*[52] the Massachusetts Supreme Judicial Court found that no regulatory taking of coastal property had occurred under the *Penn Central* analysis when the Zoning Board of Appeals denied the property owner a building permit for an undeveloped parcel of land.[53]

First, the court concluded, "the evidence clearly establishes a reasonable relationship between the prohibition against residential development on lot 93 and legitimate State interests"—namely, "potential danger to rescue workers" and the concern "that in an especially severe storm, the proposed house 'could certainly be picked up off its foundation and floated' away, potentially damaging neighboring homes."[54] Importantly, coastal storms had flooded the lot a number of times in the past (in 1938, 1944, 1954, and 1991), with the result that

> Lot 93 is a highly marginal parcel of land, *exposed to the ravages of nature*, that for good reason remained undeveloped for several decades even as more habitable properties in the vicinity were put to various productive uses. *Lot 93 is now even more vulnerable than ever to coastal flooding.* Nevertheless, recent appreciation in coastal property (belatedly, and for the time being) has given the parcel some development value.[55]

With respect to "the character of the government action," the Massachusetts Supreme Judicial Court emphasized that the building restriction

> is the type of limited protection against harmful private land use that routinely has withstood allegations of regulatory takings. . . . The judge found that "it is undisputed that [lot 93] lies in the flood plain and that its potential flooding would adversely affect the surrounding areas" if the property were developed with a house. Reasonable government action mitigating such harm, at the very least when it does not involve a "total" regulatory taking or a physical invasion, typically does not require compensation.[56]

In other words, the denial of the building permit smacked strongly of traditional government efforts to prevent or reduce the occurrence of public nuisances—problems that can arise specifically because developments in the coastal zone regularly partake of the ocean's variability and power. Coastal communities, in other words, very much include the sea and its changing dynamics.

The New Jersey Supreme Court revealed a similar sensitivity to the vulnerabilities of coastal properties in 2013 when it evaluated the compensation owed to coastal landowners for an easement that the local government (the borough of Harvey Cedars) took by eminent domain in order to construct protective coastal dunes.[57] The court's decision acknowledged that both coastal governments *and* prospective property purchasers are now acutely aware that oceanfront properties are vulnerable to coastal inundation, an awareness that affects these properties' market values despite the general attractiveness of beachfront homes. Underscoring the court's perception was the timing involved in the case: the property survived Superstorm Sandy, which hit shortly after the New Jersey Supreme Court's decision, largely because of the dune enlargement project, changing both the Karans' and their neighbors' valuation of the dunes that blocked their view.

These decisions demonstrate that at least some courts appear to be evolving the application of regulatory takings law to incorporate both the realities of the Anthropocene and the fact that coastal property is, in a very real sense, potentially part of the ocean and that the ocean is both a desired and complicating component of coastal communities.

Other courts are also beginning to invoke an explicitly communitarian perspective on real property in the face of takings claims that challenge government measures to protect public health and welfare. For example, in 2005, the Washington Court of Appeals upheld the state's destruction of privately owned citrus trees to prevent infestation of the citrus longhorned beetle—again an example that social and ecological systems are the same community—against a takings claim. As part of its decision, the court emphasized that private landowners exist within a community. Specifically, it emphasized that property law

> recognizes the reciprocal obligations of property owners to each other and to the surrounding community. The power that the State has to prohibit such uses of property as may be injurious to the health, morals, or safety of the public is not, and cannot be, "'burdened with the condition that the State must compensate such individual owners for pecuniary losses they may sustain, by reason of their not being permitted, by a noxious use of their property, to inflict injury upon the community.'"[58]

As such, the court underscored the true power of the traditional police power in takings litigation, which is to invoke with considerable legal strength the interests of the community as a whole—including the ability of those interests to override, without compensation, the narrower, short-term, and limited interests of individual private property owners.

With factors including the nature of the state interest, the extent and character of the regulatory burden, and the expectations of the private property owner, it is easy to see why this is such a complex and dynamic area of law.[59] The Takings Clauses hold within them the inherent tension between private rights and public welfare. Court interpretations place limits on property in perhaps a surprising way, given the dominant property discourse of the culture. On the whole, takings cases display a strong respect *for* and support *of* government regulation. The various "tests" and "per se rules" related to the Takings Clause, as applied, mean that a government regulation rarely actually constitutes a taking. Instead, there is an implicit recognition that we are part of a community, and constraints on the use of property are often necessary in order to implement and protect societal goals. Only in the extreme cases is the government required to pay. As such, the case law demonstrates what legal geographer Nick Blomley says has been so often the case—there is a mismatch between "property rights" as a legal form/model invoked by many property scholars and the reality of what it actually does.[60]

In all of the cases discussed (with the possible exception of the mining of coal in *Mahon*) the relational context of the properties involved is saturated with legal and regulatory processes and requirements that *create* rather than *diminish* value. The value of the property—even its existence as property—cannot be separated from its societal context. That context includes the role of the state in securing and enforcing a multitude of social contracts reflecting culturally based values and expectations. It also increasingly includes courts' recognition of Leopold's main teachings, that private property is always part of larger, complex SESs. As such, courts' evaluations of government regulation of property in the Anthropocene is likely to increasingly take account of the larger system processes of which private property is part, from coastal storms to climate change to the production of pests and disease vectors.

Conclusion: Communitarianism, Resilience Theory, and the Anthropocene

When the circumstances surrounding a community change, the community must respond *as* a community. Resilience theory provides a conceptual framework for reconfiguring our definition of "community" to recognize our inextricable connectedness with natural systems. We *are* these systems, as the realities of the Anthropocene are beginning to demonstrate. Leopold recognized that, as part of this system, humans have an ethical obligation to broaden our thinking beyond ourselves to also consider the broader land community.

We call on resilience scholars to take this ethical turn in their applications of the theory in recognition of the fact that there are no value-neutral applications of resilience. While the theory itself does not inherently value any particular system characteristic above others, the reality is that *we humans do*. This reality means that any engagement of resilience theory must take responsibility for the cultural beliefs and values reflected in what is produced. For this reason, use of terms such as "natural capital" and "ecosystem services" are problematic because they reinforce a nature-as-commodity mentality. Leopold saw the danger in this thinking, and we are living with its legacy now. The invocation of ecosystem services, natural capital, and other neoliberal ways of describing the land community is closely related to the rights-based liberalism rejected by Leopold. It is also dangerously close to the Manifest Destiny narrative. The proposed and more radical vision for resilience theory includes, of course, the trickster—the agent of change that will encourage humility and teach us to embrace the lessons that come with change.

Leopold knew something that some resilience scholars do not: all environmental challenges are also ethical challenges. As this chapter has already suggested, this new ethic should prompt a goal of maintaining the health of social-ecological *communities* as much as possible throughout the transformations that have already begun, recognizing that resilience theory cautions us that we may well lose some parts of existing SESs.

There is some evidence that the law is starting to "get it"—that we cannot continue to privilege human demands for natural resources at the expense of undermining larger-scale complex system dynamics or the future of other species. For example, the US District Court for the District of Massachusetts upheld the National Marine Fisheries Service's decision to shut down the East Coast spiny dogfish fishery for five years because it was overfished, concluding despite fishermen's protests that "as a sick person must undergo painful surgery and then convalesce for a short time in order to regain his health, a sick fishery must suffer this drastic procedure and then conserve itself for a short time in order to recover its full vitality."[61] As in Leopold's ethic, the health of the larger community—which here includes the spiny dogfish—matters.

As the above analysis of US property rights and takings jurisprudence indicates, the law continues to adapt and change to reflect new environmental realities. The law has always acknowledged that private property and its use are subject to the larger public welfare and part of a larger system. In our final chapter, we provide additional law and policy changes that, if they occur at a pace and scale worthy of the challenges we face in the Anthropocene, will promote the continued health of the larger human community despite the Anthropocene and the trickster climate change. None of these proposed changes in governance, however, will be as successful if we do not change how we perceive our role in the world. In order to *think like a system,* we must break down the false binary separating social and ecological aspects of systems at various scales.

CONCLUSION

Living the New Story

Implications for Governance

We hope that, by now, we've convinced you that the realities of the Anthropocene demand a new approach to environmental governance. This is not a conclusion to be taken lightly. Nor is it popular. Change is rarely welcome, and, when that change is characterized by uncertainty, complexity, and a need for disruption, it becomes particularly difficult to accept.[1] Climate change is not the only factor precipitating the Anthropocene, but it is the most profound. We agree with those who place the beginning of this new era with the birth of the Industrial Revolution, even though the official recommendation at the moment is to date it to the atomic bomb. With the Industrial Revolution, humans started to harness and use fossil fuels at scales that began to alter planetary climate systems and ocean chemistry. It was the true start of humanity's proven ability to influence the entire Earth.

The Industrial Revolution resulted in unprecedented human prosperity. In *Nature Unbound,* Linus Blomqvist and his colleagues Ted Nordhaus and Michael Shellenberger observe that

> global average life expectancy at birth has risen from 30 to 70 years since 1900. Between 1980 and 2010, the share of the global population living in extreme poverty dropped from 50% to 20%. On most measures of human well-being, developing countries have been converging with the developed world over the past few decades. As the global population has more than quadrupled since 1900, and global per-capita income has increased at least fivefold, total consumption of economic goods like food, water, materials, energy, and living space has vastly increased.[2]

This prosperity was made possible by cheap energy in the form of fossil fuels. The many benefits of fossil fuels should not go unrecognized as we now deal with their environmental consequences. At the same time, climate change and its consequences demand our attention.

To date, there has been a collective failure to acknowledge the realities of the Anthropocene and refigure our environmental and natural resources laws and policies accordingly. This failure can be attributed partly to the way we tend to conceptualize our relationship to the natural world. Current narratives that we use to deal with climate change, presented in Chapter 1, range from denial to nihilism, but they share common characteristics of being unhelpful and, for the most part, disempowering. This failure is not surprising given the more general narratives that Americans have employed—as outlined in Chapter 2—to describe their relationship to the environment. They include (1) Manifest Destiny and its reckless commodification of nature; (2) the tragedy paradigm of the 1970s and our early efforts to engineer our way out of the problems we created; and (3) sustainability and its idealistic notion that we can find a way to meet everyone's needs and, more importantly, desires and still have a healthy, stable natural environment.

Unfortunately, the nature and complexity of climate change and other elements of the Anthropocene do not lend themselves to easy or complete resolution or control. That is *not* to say that immediate action is not critical. It is, because we need to mitigate the severity of climate change's future impacts and adapt to the changes and transformations that will be occurring. However, future strategies cannot realistically be expected to "solve" these problems. We can no longer mitigate our way entirely out of climate change and ocean acidification. Instead, we are stuck with a changing world for the foreseeable future, and we need both to try to limit the severity of future change and to respond to the unavoidable changes as meaningfully and effectively as we can.

We are at a point in history where the ability to respond productively to continuing change *matters*. As Charles Darwin is purported to have said, "It's not the strongest of the species that survives, nor the most intelligent, but the one most responsive to change."[3] On an encouraging note, it is helpful to remember that, as Americans, we have taken on tremendous challenges before. World War II provides an extensive and largely noncontroversial example. In response to the bombing of Pearl Harbor and the needs of a four-year war effort, the federal government instituted both legal and propaganda measures. The government commandeered large segments of the US economy and redirected its productivity to war needs. Men were drafted into the armed forces, and women were propelled into the workforce. Children were encouraged to knit washcloths for soldiers. People grew more of their own food in what were known as "Victory Gardens," and the government rationed food, petroleum, and other commodities essential to the war effort. Society changed, and changed quickly, to accommodate new realities and a perceived threat to the entire nation.

Apropos of the discussion in Chapter 6, these changes were also accompanied by a sense of moral responsibility and a new rhetoric of communitarianism, particularly with respect to the rationing of food and the roles of working women. Amy Bentley emphasizes both "the communal nature of rationing itself, which required apportioning food equally among all Americans," and the need to keep all US citizens involved in and committed to the war effort, even at the expense of their individual desires and liberties:

> Preventing waste, avoiding black markets, producing food, and abiding by food rationing, however trivial they may have seemed, allowed Americans to contribute to, and feel a part of, the war effort in daily physical and communally oriented ways. . . . Rationing not only ensured a sufficient, if at times unexciting, diet for Americans but also instilled in Americans a sense of public commitment to the war, community involvement, and patriotism. By complying with rationing, conserving food, and even producing and preserving their own food, Americans increased their initial commitment, making it easier for them to support an extended and devastating world war. By taking an active part in official food campaigns, citizens also fulfilled a more communally oriented democratic obligation to their country.[4]

Food rationing was the epitome of communitarian balancing: it helped to fulfill the larger needs of the community and also ensured that individuals within that community survived. This sense of community applied to the United States as a whole, with some notable and despicable exceptions, including the internment of Americans of Japanese descent. Food hoarding and domination of the food market by the wealthy were, if not entirely eliminated, at least significantly thwarted, largely by redefining food's status as property. Moreover, the emphasis on community in World

War II carried with it a profound ethical and moral perspective, generally coalesced around the concept of patriotism.

World War II spurred continuing transformations in American society. Service in the war helped to support the civil rights claims of minorities, while women's service in the workforce helped to spur demands for gender equality. As Chapter 2 noted, these post–World War II changes would also, a few decades later, give rise to the tragedy narrative and current forms of US environmental and natural resources law and policy. We are still, in other words, living out the full ramifications of this cultural transformation, despite the fact that US values have since evolved to embrace one of the most radical incarnations of liberal individualism that the world has ever known—a liberal individualism made possible in large part by the United States' relative stability since World War II.

We are now transforming again. The Anthropocene is a time of profound change for the United States, albeit one that is moving more slowly than World War II. It is an era that, by its very name, underscores the fact that our relevant community is the entire planet. Climate change reveals that we both influence these systems and are being changed by them, which should prompt—as in World War II—both a new emphasis on the importance of community and a new ethical framework. This ethic, however, needs to be based not on loyalty to a country but rather on a recognition that our community includes nonhuman members and large complex systems.

The narrative of resilience, along with the cultural figure of the trickster, offers a more productive way of thinking about the Anthropocene and the challenges to come. We offer a purposefully radical vision of this era that is neither nihilist nor naïve. Resilience theory provides a theoretical framework predicated upon the very real complexity of and uncertainty surrounding social-ecological systems (SESs). By incorporating the trickster as a cultural component, we create an opportunity for Americans to develop a new relationship to climate change, one where change itself is an accepted part of life that people can cope with—even if the exact type or parameters of specific changes cannot be readily predicted.

Climate change will continue to bring the unexpected. We can resist that fact, or we can learn as much as we can and respond accordingly. Some responses will allow societies to continue more or less as they did before, albeit with necessary accommodations. Others will require transformation and the embrace of a new cultural and ecological identity.

Regardless of any particular narrative or way of conceptualizing so-cial-ecological relations, the way in which we perceive and therefore orient ourselves towards environmental challenges matters. The stories we tell about our situation not only assign meaning to past and current events, but also determine the options we perceive as being available to us. A resilience narrative has the potential to foster and develop the numerous strategies that are necessary to anticipate and negotiate our complex and rapidly changing world.

At its most productive (and most radical), a resilience narrative should promote active and continual engagement with complex systems, eschewing reductionist thinking or a search for technological panaceas. Rather than relying on science and technology to provide timely, "complete" answers, resilience thinking focuses on asking interesting questions that allow for the ongoing refinement of emerging understandings of continually changing SESs. Rather than identifying an institutional "fix," resilience thinking focuses on building the adaptive capacity of both humans and other members of our community—a vitally important characteristic whether the goal is to maintain a certain SES state or (as gracefully as possible) transform to a new one. The resilience narrative is grounded in understanding and responding to change. Importantly, however, it can also be a narrative of connection—humans are not separate from nature but part of a complex and dynamic social-ecological system. For these reasons, resilience theory provides a new way of thinking about our relationship to the environmental and natural resource challenges of the Anthropocene, providing a way to reconceptualize the stories we tell ourselves *about* ourselves and our place in the world.

Throughout this book we have figured *climate change* as the trickster, emphasizing the ability of humans to *respond to* tricksters. However, trickster tales also illustrate how *humans* operate as agents of change: In many respects, as is true with many types of cultural narratives, the trickster is humanity writ large and mythically. As such, trickster tales can operate as doubly corrective cultural narratives regarding our relationship to nature, priming Americans to move into the more rigorous realm of resilience thinking.

The Manifest Destiny narrative preached that natural resources (water, land, fish, minerals, trees, and so forth) are humans' for the taking and that consumption should be limited only by the places humans can go and the technologies that we can employ. But we know now that a rampant consumeristic culture, magnified by an ever-increasing human population that generally strives to live the lifestyles enjoyed in the United States and Europe, cannot survive indefinitely. We need a new narrative that can place human consumerism and consumption into a broader context, showing us how overconsumption at a global scale undermines the resilience of the very systems that made such consumption possible in the first place and pushes them toward or across ecological thresholds, especially in combination with climate change impacts. In following the Manifest Destiny sirens' call, humans emulate the various tricksters who end up trapping themselves in their own schemes to indulge their rapacious hunger—hoist by our own petards, as the Bard would have it.

The tragedy narrative, in turn, viewed humans as the snake in the garden. With humans as the despoiler, the more that we take control to return the environment to a state of nature, backing ourselves out of those natural spaces, the better. But we know now that humans are very much a part of the larger Earth system, from the land we need to live on to the food we need to eat to the phytoplankton- and plant-produced oxygen that we need to breathe. We also now know that we cannot—and in fact have never been able to—control the panarchical hierarchies of adaptive cycles in complex Earth systems. The choice is not (and has not been since at least the Industrial Revolution) between the stark binary of pristine ecosystems and despoiled human-occupied spaces. While the latter may certainly exist in the minds of many, the former no longer do. We do in fact influence the planet—but, like the trickster, our influences are too complex to be universally labeled "good" or "bad," and humans have repeatedly demonstrated a sometimes amazing capacity to make the world a better place. Trickster tales give us a much more nuanced set of stories about the range of possible roles that agents of change can play, from culture hero and bringer of light and warmth to the agent of the end of the world.

Finally, the sustainability narrative, at least as told in the United States, animates a belief that we live in a relatively stable world that we understand well enough to successfully balance environmental, economic, and social goals simultaneously—maybe even without having to trade off desires in one realm for desires in the other. Under this narrative, resources and ecosystems behave in predictable, mechanistic ways (a steady-state "Balance of Nature" view of ecology), and development as we've pursued it since the Industrial Revolution can be sustained indefinitely. But we now know that, as a scientific matter, the Balance of Nature description of complex natural systems was simply wrong. This notion of stationarity was rejected by the natural sciences and should now be rejected in our law and policy. We also now know that transformations of SESs are not only possible but also increasingly likely, and that human consumption of fossil fuels and destruction of biodiversity are pushing the entire planet toward some game-changing thresholds that we should probably avoid crossing. Like the trickster, humans make this world[5]—but as in the trickster tales, the results aren't always what we intended. As such, trickster tales and resilience theory should teach us caution and profound humility as we embark on the quest of mitigating and coping with the Anthropocene, emphasizing values of reduced consumption, cooperation, and deep respect for a broad and intimately interconnected community.

Changing the cultural narratives that Americans use to describe climate change and their relationship to nature is a critical first step in productively coping with the Anthropocene. Law and governance must be part of that cultural transformation, because law and governance are foundational components of adaptive and transformational capacity. The United States is at a critical juncture with regard to integrating resilience theory into actual structures of governance. We have an opportunity to change our story, but in order to do so, we have to let go of previous narratives. To date, there has been a reluctance to do so, mainly because, as we've noted before, many see shifting the discussion from sustainability to resilience as admitting defeat. But that is not the case. In fact, a resilience orientation in most cases will require *more* from our laws and institutions because it requires transparent discussions about what we value and the trade-offs we face.

In our two main case studies, we provided examples of the tough decisions and trade-offs that occur when governance institutions put resilience theory into practice. Chapter 4 demonstrates that New Mexico's forests are crossing an ecological threshold. The corresponding social systems, including the downstream communities of Albuquerque and Santa Fe, will also need to transform by revisiting past assumptions about water allocation and storage. Chapter 5 examines the cross-scale interactions within the ocean system. The ocean is panarchically transforming across multiple scales as a result of climate change and ocean acidification synergistically interacting with each other and with other anthropogenic stressors. Both the United States and other nations have taken important steps to set aside larger and larger areas of the ocean for protection. However, most predictions indicate that pervasive collapses of fish stocks will be occurring over the next three decades. With regard to marine fishing, the world faces a potentially determinative but extremely difficult choice: humans can phase out large-scale wild-caught fishing *now*, giving marine ecosystems their best chance to transform productively, or we can continue current consumption patterns and help drive the future ocean to the point where only jellyfish and some plankton survive.[6]

As we have noted at length, resilience theory does not itself posit a normative goal. It instead describes properties of complex systems—engineering resilience, ecological resilience, adaptive cycles, panarchy, thresholds, and transformability. Incorporating resilience theory into law and governance means not only accepting a new description of how SESs actually work (i.e., correcting the outdated scientific assumptions that underlie much of natural resources law and policy in the United States) but also consciously defining and often redefining the normative goals that those laws and governance institutions are actually trying to achieve. In other words, in the process of that incorporation, policymakers will need to answer the questions "resilience *of* what, *to* what, and *for* whom?" for specific systems in specific time frames.

There is no avoiding the fact that, when we incorporate resilience theory into law and policy, our values come into play. In Chapter 6, we encouraged resilience scholars and practitioners to embrace Leopold's land ethic, a communitarian value system that moves beyond a human-centered ethic toward a broader consideration of "community." By acknowledging the ethical implications of resilience theory in practice, we can bring more complexity to our characterization of both the social and the ecological and participate in broader policy-based discussions.

Governing in the Anthropocene

In *Learning to Die in the Anthropocene: Reflections on the End of a Civilization*, Roy Scranton states that "for humanity to survive the Anthropocene, we need to learn to live with and through the end of our current civilization. Change, risk, conflict, strife and death are the very processes of life, and we cannot avoid them. We must learn to accept and adapt."[7] It is time for our laws and policies in the United States to reflect what scientists have learned about natural systems. Governance in the Anthropocene will need to provide the necessary tools that will allow us to respond meaningfully to climate change and its associated challenges.

Building adaptive capacity and leaving options open for new trajectories will be core characteristics of environmental governance moving forward. In order to meaningfully engage the realities of social-ecological change in the Anthropocene, new policies and institutions must be developed that can accommodate radical uncertainty and continual change. As a starting point, environmental and natural resources law and policy might adopt the following general normative goal: *To build adaptive capacity within ecosystems and societies in order to adapt to climate change and associated stressors with the aim of promoting biodiversity and ecological function and, where necessary, of guiding chosen trajectories for SES transformation.*

There is, of course, a lot packed into this goal, even as general as it is. And of course, the devil is always in the details. Specific SESs and specific circumstances will always require a more nuanced approach to defining exact normative goals. Nevertheless, by starting with the reality that changes—including unexpected changes and transformations—are just part of life, resilience theory provides a narrative that can make these approaches possible. Moving forward, the key will be developing both formal and informal institutions that build our adaptive capacity while also providing enforceable, yet flexible, environmental standards. This "principled flexibility" will allow for a governance design scheme that: (1) identifies those areas of law, like pollution control requirements, that need to be strengthened in order to promote systemic resilience; (2) identifies, conversely, areas of law that govern changing systems but that are still grounded in assumptions of ecological stationarity and hence need to become more flexible; and (3) for the second set of areas, encouraging governance regimes to adopt priorities and goals for particular systems while simultaneously creating legal space for managers to respond and adapt to changing ecological and social realities.[8]

Resilience theory will require many different specific rules and standards to accommodate the wide variety of subjects that make up environmental and natural resources law. For example, pollution is almost always an anthropogenic stressor to SESs. Beyond immediate impacts, many forms of pollution can bioaccumulate, move across media (e.g., air to water and vice versa), and/or accumulate downstream or downcurrent. As discussed in Chapter 2, the existing laws for reducing existing stressors like pollution in the United States have done a good job of addressing the relatively straightforward problems. Nevertheless, much remains to be done. In application, resilience theory emphasizes the need to significantly reduce or eliminate as many of these non–climate change anthropogenic stressors as we can in order to reduce the number of perturbations to SESs that remain in desirable states. More provocatively, resilience theory may also require that American environmental law restructure environmental cost-benefit analyses and regulatory permitting/market entry thresholds to better account for the long-term impacts, the synergistic impacts, and the known and unknown unknowns of chemical interactions, such as the increasingly pervasive environmental interactions of pharmaceuticals and hormone mimickers in unstudied combinations.

In the context of natural resources management, resilience theory counsels for the across-the-board implementation of ecosystem-based management based on a *strong* precautionary principle. Moreover, this precautionary principle should now be informed by the new reality that all bets are off for ecosystems in a climate change era. Resilience theory also counsels that biodiversity protection should receive far more attention in US law and policy than it has to date. Embarrassingly, the United States remains one of the four United Nations–recognized countries (along with Andorra, South Sudan, and the Vatican) not to have ratified the UN Convention on Biodiversity, perhaps the most emblematic example of our willingness to subordinate biodiversity preservation to other societal goals—like economic development. Widespread extinction of species is consistently predicted as a climate change impact. Moreover, loss of biodiversity also impairs SESs. As discussed, the Planetary Boundaries Project identified biodiversity loss as a planetary threshold that we are at significant risk of crossing. It also identified biodiversity as one of the two most important components for the future of the planet.

The Endangered Species Act (ESA) provides an example of what reforms might be necessary in the Anthropocene in order to protect biodiversity. Passed in 1973, the ESA is one of the most powerful and controversial environmental laws in the United States. As discussed in Chapter 3, because of the law's uncompromising position against biodiversity loss, the ESA has become the primary driver of many ecological restoration efforts in the United States. In its current form, however, the law itself has limited capacity to effectively engage the complexity of SESs. Limitations include (1) the ESA's focus on individual species rather than the overall functionality of ecological systems, (2) the fact that the law begins to protect species only when federal agencies formally recognize that they are either threatened with or in danger of extinction, (3) limits on its protection to the historic range of a species, and (4) the law's outdated assumptions regarding eco-logical equilibrium and stationarity.

That said, significant reorientations of the ESA are possible. First and foremost, there is a need to shift our management strategies from a spe-cies-centered to a systems-based approach. Moving from a focus on spe-cific species or even particular habitats would allow managers to capture the complexity associated with the challenges of biodiversity loss. New approaches need to allow for the formulation of meaningful responses that foster biodiversity while also increasing our understanding of the systems states involved. The ESA, while focusing much of its protection efforts on individual species, also has the management task of protecting the ecosys-tems upon which they depend.[9] The combined efforts of the US Fish and Wildlife Service and many western states to protect the greater sage-grouse and the ecosystems it needs across most of its 173-million-acre range pro-vide a specific example of potential new approaches to biodiversity protec-tion—and, notably, these efforts are being undertaken in order to keep the grouse *off* the ESA list.[10]

A second and related issue is that building the resilience of species to cli-mate change impacts will require more proactive management efforts that support the functioning of ecosystem processes *before* they are endangered and on the brink of regime change. As noted, the ESA provides substantive protections only for species that are threatened with or in danger of extinc-tion. Other species (with some exceptions, most notably migratory birds and marine mammals) are generally the responsibility of the state in which they reside. In 2000, Congress created the State Wildlife Grants Program, which directs each state to create a Wildlife Action Plan that takes a land-scape-scale approach to biodiversity conservation.[11] This new law provides an example of multi-institutional, collaborative conservation efforts that are federally supported but system-based.[12]

Among other things, this new law requires Wildlife Action Plans to use adaptive management, a key governance tool for resilience-oriented practitioners. Adaptive management is sometimes described as "learning by doing," but, properly implemented, it requires a rigorous process of monitored experimentation in management to either elucidate some of the "known unknowns" about how a particular system functions or, in-creasingly commonly, to adapt management measures to a continually changing system.[13] However, increased deployment of adaptive manage-ment requires substantial changes in current administrative law rules, as several scholars have recognized.[14] Thus, incorporating resilience theory into environmental and natural resources law will promote "best practices," like adaptive management, that may in turn require other adaptations in different kinds of laws.[15]

Finally, a resilience reorientation to management under the ESA could move away from goals associated with preserving, restoring, and optimizing species populations and their habitats and toward goals associated with fostering complexity and adaptive capacity. Restoration is often stated as a management goal, but for most species and their associated ecosystems there is no "going back" to a time without water diversions, urban populations, and other continuing challenges—not to mention climate change's alterations to precipitation, vegetation, species assemblages, and so forth. Moving away from the restoration/preservation focus is made even more necessary when managers are confronted with the realities of the Anthropocene. For the most part, ESA implementation fails to face the reality of climate change, basing recovery efforts on the historic ranges for species and not explicitly recognizing issues related to shifting habitat distributions, migration patterns, and the need for ongoing management—although adjustments in how the act is implemented could be made.[16]

Other kinds of new tools are needed to accommodate continuing resource use while increasing biodiversity protections. One example is TurtleWatch, a real-time ocean temperature tracking system that allows longline fishermen in Hawai'i to avoid catching sea turtles, which are protected under the ESA. Sea turtles like to stay in water of a certain temperature range, and so the system effectively provides fishers with continually adjusted "no-go" zones, dramatically reducing the bycatch of these endangered and threatened species.[17]

From a governance perspective, however, improving the ESA and other existing statutes will not be enough. Any real integration of resilience theory will require a number of changes in our approach, including new laws and institutions that better equip us to face and acknowledge that regime shifts are occurring and will continue to occur.

Leopold's call for societies to wisely save the "parts" of the ecological systems is even more relevant today than it was in 1949 when *A Sand County Almanac* was published. The ESA has been an important tool in this respect. Resilience theory, however, suggests that a systems approach to biodiversity warrants a more complex, flexible, and iterative set of management tools. Just keeping the "parts" isn't enough anymore; we also need to promote their abilities to assemble and reassemble as functional ecosystems. For example, we can increase species' ability to adapt for themselves if we do two things: (1) protect as many currently useful habitats and ecosystems as we can; and (2) create and protect as many corridors as possible to connect those habitats and ecosystems so that species can move to new ranges as they need to.

Resilience theory also counsels that formerly taboo subjects need to be incorporated into US law and policy. Assisted migration for species is one example. While we can have quite a spirited debate on the "proper" role of humans in actively assisting species' survival, such as through assisted migration, our knowledge of whether such interventions will work is still fairly limited, especially in light of the fact that many species, terrestrial and marine, are now shifting their ranges to accommodate climate change impacts. Nevertheless, assisted migration may well become the only means of helping some species to survive outside of zoos and other captive breeding programs, and it needs to remain an option.

Resource consumption and human population controls are two other generally taboo topics. Nevertheless, they must be part of the discussions regarding environmental, natural resources, and energy law and policy. The current population growth rate in the United States is 1 percent per year, down from its high point of 2.1 percent in the 1970s.[18] "Population growth today is primarily driven by longer life spans and lower infant mortality, not by rising fertility."[19] In the United States:

> Between 2014 and 2060, the U.S. population is projected to increase from 319 million to 417 million, reaching 400 million in 2051. The U.S. population is projected to grow more slowly in future decades than in the recent past, as these projections assume that fertility rates will continue to decline and that there will be a modest decline in the overall rate of net international migration. By 2030, one in five Americans is projected to be 65 and over; by 2044, more than half of all Americans are projected to belong to a minority group (any group other than non-Hispanic White alone); and by 2060, nearly one in five of the nation's total population is projected to be foreign born.[20]

The rate of consumption for this growing population creates the main challenge for increasing the resilience of ecosystems and individual species to climate change. The United States ranks highest in most consumer categories by a considerable margin, even among industrial nations.[21] As noted above, trickster tales teach us that acting on strong appetites has a way of trapping the trickster. Americans would aid their own and other nations' future adaptation efforts by learning those lessons and actively discouraging conspicuous consumption.

Legal and Institutional Design— Government and Governance

The challenge becomes how to design new governance structures that thoroughly incorporate resilience theory and its expectations of unexpected change. These structures must address the need for adaptive capacity and administrative flexibility while also providing the necessary strong and enforceable frameworks that will sufficiently support the SES states that we seek to foster and protect. The tension between enforceability and flexibility and the difficulty of accommodating both within current environmental management challenges have become the focus of legal scholars paying close attention to the interrelationship of conservation science and law.[22] For example, in "General Design Principles for Resilience and Adaptive Capacity in Legal Systems: Applications to Climate Change Adaptation Law,"[23] J. B. Ruhl provides some suggestions for designing legal systems that are themselves resilient and therefore more responsive to climate change and other challenges. Noting the extent to which this design effort will require a significant departure from the status quo, Ruhl emphasizes how the current legal system is preoccupied with certainty and finality and the difficulty many federal agencies are having in incorporating adaptive management as a primary vehicle for resilience theory:

> The problem is that natural resource management agencies are locked in an administrative law system that . . . shows no sign of being flexible in that regard. The system's fixation on predecisional environmental assessment, cost-benefit analysis, records of decisions, and judicial review litigation has only pushed the system toward a "front-end" focus on reliability and efficiency that has made adaptive management exceptionally difficult to implement.[24]

Ruhl focuses on strategies for building adaptive capacity within the legal system. He identifies the need to: (1) move away from the current level of investment in land use planning, NEPA, and other processes that are inherently built on assumptions of stationarity and predictability;[25] (2) embrace strategies that are emerging from new governance theory including less emphasis on command-and-control and more encouragement of collaborative, polycentric, and adaptive models of governance;[26] (3) invoke dynamic federalism as an approach for addressing the multi-scalar dimension of climate change and other challenges;[27] and (4) encourage the formation and maintenance of transgovernmental networks as informal but critical linkages across scales of governance that promote information sharing and social learning.[28]

Flexibility and adaptive capacity will be important moving forward, but so will changes in our use of the rule of law. Beyond redesigning administrative law to accommodate adaptive management and other flexible management procedures,[29] the law needs to incorporate new designs that allow for flexibility without turning natural resources management into an agency free-for-all. Principled flexibility requires designing and implementing environmental policies that promote and build adaptive capacity to respond to changing environmental conditions while also providing stronger, more legally enforceable and institutionally supported goals to reduce existing and preventable stressors on SESs, increasing their resilience to climate change impacts.[30]

In its current stage of integration and development, resilience is in danger of becoming—like sustainability—a rhetorical device with little influence on actual decision making. We are at a critical point with regard to the challenge of integrating resilience thinking into environmental policies and approaches. Our focus has been on adaptation strategies necessary to enhance our capacity to deal with the change ahead. In "'Stationarity Is Dead'—Long Live Transformation,"[31] several basic principles are provided to build this capacity. We restate the relevant recommendations here in order to provide some specific ideas for more comprehensive integration of resilience theory into environmental and natural resources law and policy in the United States.

Monitor and Study Everything All of the Time

The first principle is to *monitor and study everything all of the time*.[32] We need to be better informed in order to be better prepared. While climate projection models continue to become more refined, there is still a high level of uncertainty regarding ongoing impacts, especially at the local level.

However, in general, ecological monitoring in the United States takes place only where there is an explicit legal mandate.[33] This fact needs to change. New climate change adaptation laws should increase requirements and funding for continual monitoring and basic scientific and economic research to promote understanding of climate change impacts at all scales and across sectors.

Almost all monitoring currently authorized by law is concerned with *known* unknowns—i.e., things that we *know* we need to learn more about. Former US secretary of defense Donald Rumsfeld is famous for pointing out the challenges of unknown unknowns—things we don't realize we need to understand but are important nonetheless.[34] These unknown unknowns, especially given the Anthropocene, are why long-term research designed to increase our understanding of SESs is so important. The Anthropocene demands a new relationship with information, one that recognizes the continually evolving nature of our understanding of SESs. Major investments in research and monitoring, particularly in the area of system threshold identification, will help overcome political impediments to identifying and implementing adaptation measures and help to prioritize legal and policy goals by clarifying which systems and system components are being most affected by climate change and its impacts, and in what ways.

Eliminate or Reduce Non–Climate Change Stresses to Promote Adaptive Capacity

Principle 2 is to *eliminate or reduce non–climate change stresses to promote adaptive capacity.* This principle forms the basis of a new and necessary "no regrets" policy.[35] Whether an SES can maintain its basic identity or undergoes transformation to a new one, reducing stressors on the system is one of the best ways to enhance a system's adaptive capacity and ecological resilience to other stressors, like climate change. Stressors come from a range of sources, not all of which are directly linked to climate change.

This principle has several subparts. The first is to *decontaminate land, water, and air, as well as reduce new pollution as much as possible.*[36] Again, the point is to build as much adaptive capacity and resilience to other stressors within the system as possible. For too long, we have allowed pollution from various sectors (like agriculture) to continue, even where the technology to address it is available. In addition, many pollution standards are linked to "best available" or "best conventional" technology, which do not encourage the necessary innovation needed in the Anthropocene.

Standards that drive improvement are called "technology forcing" because they encourage the development of new ideas. An example can be found in fuel efficiency standards for vehicles. Increased efficiency standards drive industry to do better.

The second subprinciple is to *rethink the traditional "maximum sustainable yield" regulatory standards.*[37] As noted in the case studies in Chapters 4 and 5, these standards now hold little meaning. When all components of ecosystems and higher-scale planetary systems are continually changing, we have no way of calculating any aspect of what "sustainable" use might be—and the impacts of *any* human use of resources will themselves be changing over time. Pursuing "maximum" use under these realities will serve only to drive ecosystems toward ecological thresholds and transformations—changes in ecosystem productivity and function that human societies generally will not want.

The third subprinciple is to *stop subsidizing/encouraging maladaptive behaviors and provide incentives for adaptive behaviors.*[38] Incentive structures are a long-standing environmental management tool. Unfortunately, law sometimes creates perverse incentives that actually discourage innovation. For example, the Clean Air Act has rigorous pollution standards for new coal-fired power plants. These stringent emissions standards also apply if an existing coal-fired power plant undergoes significant modifications or upgrades, creating one of the act's most insidious (if unintended) perverse incentives: Industries have *avoided* upgrading or replacing their facilities to avoid the more vigorous standards. As a result, more than half—51 percent—of the United States' electricity-generating capacity was built before 1980, and about 74 percent of all coal-fired power plants are at least thirty years old.[39] These power plants generally still emit pollutants at the old polluting rates. Congress and the Environmental Protection Agency (EPA) should fix this perverse incentive. At the same time, the United States needs to end its subsidies for fossil fuel development and other climate change–inducing activities.

Finally, this principle of reducing stressors includes the need to preserve and expand open space and ecosystem connectivity.[40] Habitat fragmentation is a primary driver of species extinction. In the Anthropocene, climate change is likely to outstrip, or at the very least severely challenge, species' and ecosystems' intrinsic capacities to adapt, even if those capacities are not already diminished by anthropogenic stressors. One of the most effective adaptation measures humans could implement may be to preserve as much connected and varied open space as is physically and politically possible to allow species and ecosystems to respond as they will to climate change impacts.

Plan for the Long Term with Increased Coordination across Sectors, Interests, and Government

Climate change will require new additional institutional capacity at all levels of government. Beyond "government," informal institutions and networks will be a necessary component moving forward. Climate change is a multi-scalar challenge and will require extensive coordination and planning. The first two necessary steps for this increased coordination are to acknowledge that climate change exists at all levels of governmental planning and to acknowledge the potential for conflicts among adaptation efforts at various scales or for various purposes and between adaptation and mitigation strategies.[41]

Because of the inherent uncertainties involved with both climate change and system responses, however, the third subprinciple is to *consider a range of possible long-term futures when planning*.[42] This exhortation invokes an important aspect of systems thinking: scenario planning. Scenario planning is a systemic method for thinking creatively about possible complex and uncertain futures. The central idea is to consider a variety of possible futures that include many of the important uncertainties in the system rather than to focus on the accurate prediction of a single outcome.[43]

The US Department of the Interior's approach to decision making with the level of uncertainty and the level of controllability provides an example by placing these two factors on separate axes. Situations where there is little uncertainty and a high amount of control allow for "optimal control." Examples include the many pollution emissions and discharges now controlled under the Clean Air Act and Clean Water Act. These types of challenges were fairly effectively addressed with the "engineering" approach described in Chapter 2 during the beginning of the regulatory era—although, as noted, both statutes need to be expanded to deal with multi-media and synergistic pollution issues.

Adaptive management is helpful when there is a high level of uncertainty, but only if managers have a high level of control over the system so that they can make changes to reflect what is learned.[44] Interior uses the term "hedging" to describe what you can do in the face of low uncertainty but a significant lack of control, a form of "no regrets" thinking. If you keep a three-day supply of food and water in your house for use in case of an emergency, for example, you are hedging. In the environmental management context, seed banking is an example of hedging. Around the world, botanical gardens and other programs are collecting and conserving native plant species to protect against disease or a catastrophic loss of genetic diversity. In these examples, the potential disasters are predictable— and in many cases (like hurricanes in the Southeast) virtually certain to occur at some point—but impossible for humans to prevent or control.

However, in situations where there is both a high level of uncertainty about future events and little capacity to control them, scenario planning can be a worthy investment. For most locations, climate change is one of these situations, particularly when it comes to long-term planning. Climate change is happening. We have little ability to control this fact (although, again, climate change mitigation efforts remain critical). We also have great uncertainty about what exactly climate change will mean for governance issues, particularly at smaller scales and more than a decade into the future. Nevertheless, we know enough to start creating various future climate scenarios and identifying our options for adapting. Scenario planning is a method for thinking systematically about and understanding the nature and impact of the most uncertain and important driving forces influencing the SES. Decision makers need to work to continually refine their understanding of the system and its dynamics in order to make decisions in the face of continual change.

The "decision makers" looking at these various scenarios need to include a wide variety of stakeholders and institutions, at a variety of scales, including global-scale coalitions of governments, corporations, and nongovernmental organizations. Moreover, decision makers should be looking first to identify "no regrets" options—that is, adaptation strategies that are a good idea for almost all scales and constituents and in both the present and the foreseeable future, regardless of climate change impacts. Efforts to reduce pollution and waste generally fall into this category, as may efforts to reduce consumerism and natural resource exploitation. The second goal of scenario planning is to try to identify potential conflicts in adaptation strategies, including: (1) adaptation efforts that may be attractive in the short term but undesirable in the long term (coastal armoring will generally fall into this category); (2) adaptation efforts that are attractive at one scale or location but disastrous at another (increasing efforts to move fresh water around the country probably falls into this category); or (3) adaptation options that might be good for humans but disastrous for other species and ecosystems (massive increases in coastal desalination would probably fit into this category, as, again, would massive new water transfers). Finally, scenario planning should be used to look for ways to avoid complicating adaptation efforts in the future or limiting future options. Coastal communities, for example, should probably avoid building critical and long-lived new infrastructure in the immediate coastal zone, where it can become increasingly vulnerable to sea-level rise and coastal storms while simultaneously impeding coastal retreat efforts. Similarly, housing developments in the urban-wildland interface are more susceptible to wildfire, potentially increasing future vulnerabilities. Many communities may discover that containing suburban sprawl, particularly by limiting expansion into floodplains and forests, will provide them more flexibility for the future while simultaneously preserving open spaces and corridors that other species can use to adapt. Communities experiencing water supply stress for the first time may choose to impose minimum stream flow requirements and aquifer protections *before* these resources become over-exploited, again benefiting not just the human but the greater ecological community while increasing options for the future.

Give Meaningful Weight to Public Rights and Values in Private Property

In Chapter 6, we explained that private property has always been subject to limits based on the needs and values of the larger community. Nevertheless, in the United States, one source of resistance to significant adaptation measures is likely to be popular conceptions of private property rights as "absolute."[45] Fear of constitutional "takings" liability is also likely to inspire at least some governments to drag their proverbial feet in implementing necessary measures.[46] To aid the incorporation of resilience theory into environmental and natural resources law and policy, law and policy need to reinvigorate and potentially expand the many legal doctrines that acknowledge public rights and values in property and ecosystems—doctrines such as the public trust doctrine, the public necessity doctrine, and public nuisance principles.[47]

As discussed in Chapter 6, constitutional takings litigation, especially along the coast, is starting to take greater account of public values with respect to vulnerable properties, a trend that is likely only to increase as sea-level rise and severe coastal storms become more pronounced. It is important to note, however, that courts have *always* displayed that same impulse to protect public values when human health is at stake.[48] When lawmakers, policymakers, managers, and the public at large better understand the real threats to health—human and ecological—that the Anthropocene poses, it will be but a short legal step to privilege a broader conception of "public health" over private interests. This is but one example of how climate change adaptation law should anticipate several alterations in cultural norms—much as World War II required.

Promote Principled Flexibility in Regulatory Goals and Natural Resource Management

Changing environmental and natural resources law and policies in the United States into mechanisms for principled flexibility requires some fine-tuning of our existing statutes. First, as part of their pervasive incorporation of restoration goals, these laws tend to assume a pre-European baseline of what ecosystems and species populations "should" be. Given European settlement, recovering those baseline states has probably always been impossible (politically if not technologically), but climate change makes any attempts to reach these baselines an increasingly futile goal. Instead of pursuing restoration to a pre-settlement state (or as close as is thought possible), environmental and natural resources laws need instead to acknowledge that SESs are changing. For example, the federal Clean Water Act mandates that uses of waterways existing at the time the act took effect (early 1970s) *must* continue to exist, forever.[49] The original goal of these provisions was to ensure that waters in the United States do not degrade any further. However, climate change is already making some of these "existing uses" impossible to maintain, such as by warming streams so much that they can no longer support cold-water species like salmon and trout. States unable to meet baseline conditions because of climate change need a regulatory method for acknowledging changes in water quality that are beyond their control while simultaneously remaining bound by the original goal of preventing increased pollution and degradation that *can* still be controlled.

Adaptive management, as noted, is an important tool for incorporating resilience theory into natural resources management practice. As already noted, agencies are increasingly required to engage in adaptive management. In order to be successful, however, legislatures need to amend related legal requirements so agencies can implement scientifically valid adaptive management, instead of using what J. B. Ruhl and Robert Fischman have termed "a/m lite"—a "watered-down version of the theory that resembles ad hoc contingency planning more than it does planned learning while doing."[50]

Finally, the "flexibility" part of principled flexibility includes preserving (to the extent possible) options for a changing future. As noted, some adaptation strategies pursued today in light of current priorities (say, preserving the status quo for as long as possible in coastal cities) may constrain or eliminate the options available to future incarnations of that SES. In addition to engaging in scenario planning, governance entities should adopt policies that prefer "no regrets" adaptation strategies over all others. They should also adopt policies that require decision makers to make future-appropriate decisions about expensive and/or long-lasting adaptation strategies. There is an understandable tendency to want to "do something" quickly and adopt simple "solutions" or panaceas that cannot reflect the true complexities of the problem, then consider the problem "resolved." Failure to acknowledge the complexity and changing understanding of climate change impacts, however, will not lead to effective climate change adaptation strategies at any level. Instead, decision makers should be cognizant that retaining as much flexibility as possible is itself an important adaptation strategy. Climate change adaptation law should require robust decision-making processes that identify adaptation measures that will be helpful under a variety of climate change scenarios for many adaptation decisions.

Conclusion

When Roy Scranton suggests that we learn "to die in the Anthropocene," he means that we need to die in the past, leaving behind what we used to know, the world as it was.[51] We join Scranton and others who stress the importance of a new cultural narrative. As he notes, "In order for us to adapt to this strange new world, we're going to need more than scientific reports and military policy. We're going to need new ideas. We're going to need new myths and stories, a new conceptual understanding of reality."[52]

In this book, we share our own sense of what these new cultural narratives should be. We offer resilience theory as a conceptual tool for thinking about the dynamics of SESs and as a more productive orientation to the

Anthropocene. It provides a science-based framework for acknowledging that complex SESs are continually changing. Climate change acts to disturb and accelerate those dynamics, making the possibility of system transformation more likely. We offer the trickster as a needed addition to American culture. These stories constitute a cultural/mythological narrative that acknowledges the recurring role of unexpected change in SESs as well as human capacity to respond to those changes. It also teaches us that change comes in many forms and that rampant consumption is often far less productive than negotiated sharing.

Climate change is the trickster of the Anthropocene. It guarantees that there will be challenges for both SESs in general and environmental and natural resources law in particular. However, coping with these challenges may also bring wisdom and a better understanding of how humans participate in, rather than control, the multiplicity of complex systems of which we are a part.

We encourage resilience thinkers to embrace a normative vision based on a narrative of connection and community, an empowering narrative of limited but influential agency that leaves humans the capacity to be a responsible member of the future Planet Earth. We must create a new story that emphasizes the relationality that resilience theory makes so clear, a story informed by a communitarian ethic along the lines of that offered by Aldo Leopold. The world is indeed changing, but we are, in the most profound sense, all in this together.

Notes

Preface

1. Robin Kundis Craig, "'Stationarity Is Dead'—Long Live Transformation: Five Principles for Climate Change Adaptation Law," *Harvard Environmental Law Review* 34 (2010): 9–73.

2. Robin Kundis Craig and Melinda Harm Benson, "Replacing Sustainability," *Akron Law Review* 46, no. 4 (2013): 841–880; Melinda Harm Benson and Robin Kundis Craig, "The End of Sustainability," *Society & Natural Resources: An International Journal* 27 (2014): 777–782.

3. Melinda Harm Benson and Robin K. Craig, "The End of Sustainability," *Ensia*, July 8, 2014, https://ensia.com/voices/the-end-of-sustainability//.

Chapter 1. Welcome to the Anthropocene

Parts of this chapter were developed earlier in: Robin Kundis Craig, "Learning to Live with the Trickster: Narrating Climate Change and the Value of Resilience Thinking." *Pace Environmental Law Review* 33 (Spring 2016): 351–396; Melinda Harm Benson and Robin Kundis Craig, "The End of Sustainability," *Society & Natural Resources: An International Journal* 27 (2014): 777–782, DOI:10.1080/089 41920.2014.901467; Barbara Cosens, Lance Gunderson, Craig Allen, and Melinda Harm Benson, "Identifying Legal, Ecological and Governance Obstacles, and Opportunities for Adapting to Climate Change," *Sustainability* 6 (2014): 2338–2356; Michael Burger, Elizabeth Burleson, Rebecca M. Bratspies, Robin Kundis Craig, David M. Crisen, Alexandra R. Harrington, Keith H. Hirokawa, Sarah Krakoff, Katrina Fischer Kuh, Stephen R. Miller, Jessica Owley, Patrick Parenteau, Melissa Powers, Shannon M. Roesler, and Jonathan Rosenbloom, "Rethinking Sustainability to Meet the Climate Change Challenge," *Environmental Law Reporter* 43 (2013): 10342–10357; Robin Kundis Craig and Melinda Harm Benson, "Replacing Sustainability," *Akron Law Review* 46, no. 4 (2013): 841–880. Use of these previous works either conforms with the original copyright or is with permission of the publisher.

1. The IPCC's Fifth Assessment Report consists of four reports published in 2013 and 2014: *Climate Change 2013: The Physical Science Basis* (2013) [hereinafter 2013 IPCC Physical Science Report]; *Climate Change 2014: Impacts, Adaptation, and Vulnerability* (2014) [hereinafter 2014 IPCC Adaptation Report]; *Climate Change*

2014: Mitigation of Climate Change (2014); and *Climate Change 2014: Synthesis Report* (2014) [hereinafter 2014 IPCC Synthesis Report], all available at http://www.ipcc.ch. Numerous national and regional reports have documented climate impacts on more local scales. US Global Change Research Program, "Highlights of Climate Change Impacts in the United States: The Third National Climate Assessment" (2014), http://www.globalchange.gov/browse/reports/highlights-climate-change-impacts-united-states-third-national-climate-assessment.

2. 2014 IPCC Adaptation Report, 2, http://www.ipcc.ch/report/ar5/wg2/.

3. Robin Kundis Craig, "'Stationarity Is Dead'—Long Live Transformation: Five Principles for Climate Change Adaptation Law," *Harvard Environmental Law Review* 34 (2010): 32–37, http://www.law.harvard.edu/students/orgs/elr/vol34_1/9-74.pdf.

4. E.g., Paul Robbins and Sarah A. Moore, "Ecological Anxiety Disorder: Diagnosing the Politics of the Anthropocene," *Cultural Geographies* 20 (2012): 3, 5, doi:10.1177/1474474012469887.

5. Damian Carrington, "The Anthropocene Epoch: Scientists Declare Dawn of Human-Influenced Age," *The Guardian,* August 29, 2016, https://www.theguardian.com/environment/2016/aug/29/declare-anthropocene-epoch-experts-urge-geological-congress-human-impact-earth.

6. Joseph Stromberg, "What Is the Anthropocene and Are We in It?," *Smithsonian Magazine,* http://www.smithsonianmag.com/science-nature/what-is-the-anthropocene-and-are-we-in-it-164801414/?no-ist (January 2013); "What Is the Anthropocene?—Current Definition and Status," Working Group on the "Anthropocene," Subcommission on Quarternary Stratigraphy, as updated May 5, 2015, http://quaternary.stratigraphy.org/workinggroups/anthropocene/.

7. "What Is the Anthropocene?—Current Definition and Status," Working Group on the "Anthropocene," Subcommission on Quarternary Stratigraphy, as updated February 23, 2016, http://quaternary.stratigraphy.org/workinggroups/anthropocene/.

8. Erle C. Ellis, Dorian Q. Fuller, Jed O. Kaplan, and Wayne G. Lutters, "Dating the Anthropocene: Towards an Empirical Global History of Human Transformation of the Terrestrial Biosphere," *ELEMENTA: Science of the Anthropocene* (2013): 1, doi:10.12952/journal.elementa.000018; Jan Zalasiewicz, Mark Williams, Alan Haywood, and Michael Ellis, "The Anthropocene: A New Epoch of Geological Time?," *Philosophical Transactions of the Royal Society A: Mathematical, Physical & Engineering Sciences* (2011): 835–841, doi:10.1098/rsta.2010.0339; Carol P. Harden et al., "Understanding Human–Landscape Interactions in the 'Anthropocene,'" *Environmental Management* 53 (2014): doi:10.1007/s00267-013-0082-0; Nicholas A. Robinson, "Keynote: Sustaining Society in the Anthropocene Epoch," *Denver Journal of International Law and Policy* 41 (2013): 467, http://digitalcommons.pace.edu/lawfaculty/927/.

9. Robbins and Moore, "Ecological Anxiety Disorder," 3, 8.

10. Ibid.

11. Ibid., 3, 9.

12. Jens Brockmeier, "Remembering and Forgetting: Narrative as Cultural Memory," *Culture & Psychology* 8 (2002): 15–43, doi:10.1177/1354067x02008001617.

13. Fritz Heider and Marianne Simmel, "An Experimental Study of Apparent Behavior," *American Journal of Psychology* 57 (1944): 243–259, doi:10.2307/1416950.

14. Richard J. Lazarus, "Super Wicked Problems and Climate Change: Restraining the Present to Liberate the Future," *Cornell Law Review* 94 (2009): 1153–1233, http://www.lawschool.cornell.edu/research/cornell-law-review/upload/Lazarus.pdf.

15. Katrin Starcke and Matthias Brand, "Decision Making under Stress: A Selective Review," *Neuroscience & Biobehavioral Reviews* 36 (April 2012): 1228–1248, doi:10.1016/j.neubiorev.2012.02.003.

16. 2014 IPCC Synthesis Report, 13; see also ibid., 16 ("The risks of abrupt or irreversible changes increase as the magnitude of the warming increases").

17. Ibid., 13.

18. Ibid., 16.

19. Ibid.

20. Ibid.

21. Ibid., 8.

22. For example, the IPCC concludes that "substantial emissions reductions over the next few decades can reduce climate risks in the 21st century and beyond, increase prospects for effective adaptation, reduce the costs and challenges of mitigation in the longer term and contribute to climate-resilient pathways for sustainable development." Ibid., 17. See also ibid., 17–19 (discussing the benefits of climate change mitigation); 20–26 (discussing mitigation pathways); and 28–29 (discussing response options for mitigation).

23. Ibid., 17.

24. Ibid., 19. See also ibid., 26 (noting the cultural aspects of effective adaptation).

25. "Narrative frames often organize our perceptions and interpretations of experience," transforming "what would otherwise be a meaningless aspect of the scene into something that is meaningful. . . . They are important guides for understanding phenomena such as social and environmental change, which develop as plots, with causal chains, perpetrators, victims, conflicts and resolutions." Thomas F. Thornton and Patricia M. Thornton, "The Mutable, the Mythical, and the Managerial: The Mutable, the Mythical, and the Managerial," *Environment and Society* 6 (2015): 66–86, doi:http://dx.doi.org/10.3167/ares.2015.060105 (quoting Erving Goffman, *Frame Analysis: An Essay on the Organization of Experience* [New York: Harper & Row, 1974]: 21).

26. Andrew J. Hoffman, *How Culture Shapes the Climate Change Debate* (Stanford, CA: Stanford University Press, 2015), 2–3.

27. Anthony Leiserowitz et al., "Global Warming's Six Americas," *Yale Project on Climate Change Communication*, September 2012, http://environment.yale .edu/climate-communication-OFF/files/Six-Americas-September-2012.pdf.

28. Hoffman, *How Culture Shapes the Climate Change Debate*, 9.

29. Brandon Baker, "Survey Shows Americans Lead the World in Climate Denial," *EcoWatch*, July 22, 2014, http://ecowatch.com/2014/07/22/americans-lead -world-climate-denial/.

30. Ibid.

31. Ibid.

32. Lydia Saad, "A Steady 57% in U.S. Blame Humans for Global Warming," *Gallup*, March 18, 2014, http://www.gallup.com/poll/167972/steady-blame-humans -global-warming.aspx.

33. Ibid.

34. "Interview: Dr. S. Fred Singer," *PBS Nova: What's Up with the Weather* (2000), http://www.pbs.org/wgbh/warming/debate/singer.html.

35. Katrina Kuh, "Agnostic Adaptation," in *Contemporary Issues in Climate Change Law and Policy: Essays Inspired by the IPCC*, Robin Kundis Craig and Stephen R. Miller, eds. (Washington, DC: Environmental Law Institute, 2016): 167–178.

36. Jeroen C. J. H. Aerts, W. J. Wouter Botzen, Kerry Emanuel, Ning Lin, Hans de Moel, and Erwann O. Michel-Kerjan, "Evaluating Flood Resilience Strategies for Coastal Megacities," *Science* 344 (May 2, 2014): 473–475, doi:10.1126 /science.1248222.

37. Ibid., 474.

38. Thornton and Thornton, "The Mutable, the Mythical, and the Managerial," 66, 67, 72.

39. Eric Niller, "Can New Energy Technology Save the Planet?," *Seeker.com*, December 1, 2015, http://news.discovery.com/tech/alternative-power-sources/can -new-energy-technology-save-the-planet-151201.htm.

40. Erle C. Ellis, "Overpopulation Is Not the Problem," *New York Times* (September 13, 2013), http://nyti.ms/18i8E5S.

41. "What Is Geoengineering?" *Oxford Geoengineering Programme*, accessed January 17, 2015, http://www.geoengineering.ox.ac.uk/what-is-geoengineering/what -is-geoengineering/.

42. Andrew Snyder-Beattie, "Geoengineering Is Fast and Cheap but Not Key to Halting Climate Change," *The Guardian*, May 15, 2015, http://www.theguard ian.com/sustainable-business/2015/may/15/geoengineering-climate-change-green house-gases.

43. John Vidal, "Geoengineering Side Effects Could Be Potentially Disastrous,

Research Shows," *The Guardian*, February 25, 2014, http://www.theguardian.com /environment/2014/feb/25/geoengineering-side-effects-potentially-disastrous-sci entists.

44. For example, "One category of geoengineering schemes, solar radiation management, has the potential to cool the atmosphere quickly and at relatively low direct cost, yet may be highly risky." Sabine Mathesius, Matthias Hofmann, Ken Caldeira, and Hans Joachim Schellnhuber, "Long-term Response of Oceans to CO_2 Removal from the Atmosphere," *Nature Climate Change* 5 (2015): 1107, doi:10.1038/nclimate2729.

45. Robin Kundis Craig, "Dealing with Ocean Acidification: The Problem, the Clean Water Act, and State and Regional Approaches," *Washington Law Review* 90 (2016): 1583–1657; University of Utah College of Law Research Paper No. 127, https://ssrn.com/abstract=2633051.

46. Mathesius, Hofmann, Caldeira, and Schellnhuber, "Long-term Response of Oceans to CO_2 Removal from the Atmosphere," 1112.

47. Tim Radford, "Stop Burning Fossil Fuels Now: There Is No CO_2 'Tech-nofix,' Scientists Warn," *The Guardian*, August 03, 2015, http://www.theguardian .com/environment/2015/aug/03/stop-burning-fossil-fuels-now-no-co2-technofix -climate-change-oceans.

48. E.g., Mark Elliott, Andrew Armstrong, Joseph Lobuglio, and Jamie Bar-tram, "Technologies for Climate Change Adaptation: The Water Sector," TNA Guidebook series (Roskilde, Denmark: United Nations Environment Programme Risoe Centre, 2011), http://orbit.dtu.dk/files/7689720/TNA_Guidebook_Adap tationWater.pdf.

49. Benjamin Gutierrez, Nathaniel Plant, and E. Robert Thieler, "Sea-level Rise Hazards and Decision Support, Coastal Groundwater Systems," *U.S. Geological Survey*, accessed November 24, 2014, http://wh.er.usgs.gov/slr/coastalgroundwater .html; Environmental Protection Agency, "Water Resources: Climate Impacts on Water Resources," November 4, 2015, http://www3.epa.gov/climatechange /impacts/water.html (accessed at this website during the Obama administration; this page now redirects to a page that states, "currently updating our website to reflect EPA's priorities under the leadership of President Trump and Administrator Pruitt," with a link to an archived location: https://19january2017snapshot.epa.gov /climatechange_.html).

50. Molly Loughney Melius and Margaret R. Caldwell, "2015 California Coastal Armoring Report: Managing Coastal Armoring and Climate Change Ad-aptation in the 21st Century," *Stanford Law School Environment and Natural Re-sources Law & Policy Program* (2015), http://law.stanford.edu/wp-content/uploads /2015/07/CalCoastArmor-FULL-REPORT-6.17.15.pdf (detailing how much of the California coast has been armored and the detrimental impacts on beaches and coastal ecosystems); Evan Lehmann. "Sea Walls May Be Cheaper Than Rising

Waters," *Scientific American ClimateWire*, February 04, 2014, http://www.scientificamerican.com/article/sea-walls-may-be-cheaper-than-rising-waters/.

51. E.g., Nick Stockton, "Map Shows Where Sea Level Rise Will Drown American Cities," *Wired.com* (October 12, 2015), http://www.wired.com/2015/10/map-shows-sea-level-rise-will-drown-american-cities/.

52. Mathew Barrett Gross and Mel Gilles, *The Last Myth: What the Rise of Apocalyptic Thinking Tells Us about America* (Amherst, NY: Prometheus Books, 2012), 9.

53. Ibid., 12.

54. Ibid., 15.

55. Michael Burger, "Environmental Law/Environmental Literature," *Ecology Law Quarterly* 40 (2013): 20, http://dx.doi.org/doi:10.15779/Z386G42.

56. Emma Green, "Half of Americans Think Climate Change Is a Sign of the Apocalypse," *The Atlantic*, November 22, 2014, http://www.theatlantic.com/politics/archive/2014/11/half-of-americans-think-climate-change-is-a-sign-of-the-apocalypse/383029/; Ryan Koronowski, "Most White Evangelicals Attribute Intense National Disasters to the Apocalypse, Not Climate Change," *Climate Progress* (November 22, 2014), http://thinkprogress.org/climate/2014/11/22/3596041/poll-religion-climate-end-times-evangelicals/.

57. James Gerken. "Climate Change Study: Religious Belief in Second Coming of Christ Could Slow Global Warming Action," *Huffington Post Australia*, May 4, 2013, http://www.huffingtonpost.com.au/2013/05/03/climate-change-study_n_3204054.html?ir=Australia.

58. Burger, "Environmental Law/Environmental Literature," 20.

59. "Mutually Assured Destruction," NuclearFiles.org, accessed January 17, 2016, http://www.nuclearfiles.org/menu/key-issues/nuclear-weapons/history/cold-war/strategy/strategy-mutual-assured-destruction.htm.

60. See Gross and Gilles, *The Last Myth*, 116: "This nuclear fear is what has distinguished the apocalyptic imagination in the modern era from all history, pushing the apocalypse from the realm of religion into the secular mainstream—a visceral shadow that has lingered at the edges of the modern imagination. . . . To a degree that the generations who passed before 1945 could never understand, the nuclear age has made our expectation of apocalypse more visceral and universal than ever before."

61. Green, "Half of Americans Think Climate Change Is a Sign of the Apocalypse."

62. Gross and Gilles, *The Last Myth*, 102.

63. Ian Joughin, Benjamin E. Smith, and Brooke Medley, "Marine Ice Sheet Collapse Potentially Underway for the Thwaites Glacier Basin, West Antarctica," *Science* 344 (May 16, 2014): 735–738, doi:10.1126/science.1249055.

64. Tim Worstall, "If Antarctic Melting Has Passed the Point of No Return, We Should Do Less about Climate Change, Not More," *Forbes,* May 13, 2014, http://www.forbes.com/sites/timworstall/2014/05/13/if-antarctic-melting-has -passed-the-point-of-no-return-we-should-do-less-about-climate-change-not -more/#2715e4857a0b3763abcb731f.

65. Thornton and Thornton, "The Mutable, the Mythical, and the Managerial," 72.

66. Eric Mack, "Melting Antarctica Is the End of the World as We Know It, and That's a Good Thing," *Forbes,* May 14, 2014, http://www.forbes.com/sites/eric mack/2014/05/14/melting-antarctica-is-the-end-of-the-world-as-we-know-it-and -thats-a-good-thing/#50c1bf723f72.

67. Thornton and Thornton, "The Mutable, the Mythical, and the Managerial," 67 (quoting Susanne C. Moser and Lisa Dilling, "Communicating Climate Change: Closing the Science-Action Gap," in John S. Dryzek, Richard B. Norgaard, and David Schlosberg, eds., *The Oxford Handbook of Climate Change and Society* (Oxford: Oxford University Press, 2011): 161–174, doi:10.1093/oxfordhb /9780199566600.003.0011.

68. Gross and Gilles, *The Last Myth,* 131.

69. "Highlights of Climate Change Impacts in the United States: The Third National Climate Assessment," US Global Change Research Program, Global Change.gov (2014): 7, http://www.globalchange.gov/browse/reports/highlights -climate-change-impacts-united-states-third-national-climate-assessment.

Chapter 2. Narrating Our Relationship with Nature

Parts of this chapter were developed earlier in: Robin Kundis Craig, "Becoming Landsick: Rethinking Sustainability in an Age of Continuous, Visible, and Irreversible Change," in Jessica Owley and Keith Hirokawa, eds., *Rethinking Sustainable Development to Meet the Climate Change Challenge* (Washington, DC: Environmental Law Institute, 2015), 41–92; Robin Kundis Craig, "Learning to Live with the Trickster: Narrating Climate Change and the Value of Resilience Thinking," *Pace Environmental Law Review* 33 (Spring 2016): 351–396; Melinda Harm Benson, "Reconceptualizing Social-Ecological Relations—Is Resilience the New Narrative?," *Journal of Environmental and Sustainability Law* 21 (2015): 99–127; Melinda Harm Benson and Robin Kundis Craig, "The End of Sustainability," *Society & Natural Resources: An International Journal* 27 (2014): 777–782, DOI:10.1080/08941920.2014.901467; Michael Burger, Elizabeth Burleson, Rebecca M. Bratspies, Robin Kundis Craig, David M. Crisen, Alexandra R. Harrington, Keith H. Hirokawa, Sarah Krakoff, Katrina Fischer Kuh, Stephen R. Miller, Jessica Owley, Patrick Parenteau, Melissa Powers, Shannon M. Roesler, and Jonathan Rosenbloom, "Rethinking Sustainability to Meet the Climate Change

Challenge," *Environmental Law Reporter* 43 (2013): 10342–10357; Robin Kundis Craig and Melinda Harm Benson, "Replacing Sustainability," *Akron Law Review* 46, no. 4 (2013): 841–880. Use of these previous works either conforms with the original copyright or is with permission of the publisher.

1. Christine Metteer Lorillard, "Stories That Make the Law Free: Literature as a Bridge Between the Law and the Culture in Which It Must Exist," *Texas Wesleyan Law Review* 12 (2005): 251.

2. Jeffrey D. Jackson, "For Effective Persuasion, Don't Neglect the Narrative," *Journal of the Kansas Bar Association* 84 (April 2015): 12.

3. Randy Gordon, *Rehumanizing Law: A Narrative Theory of Law and Democracy* (Toronto: University of Toronto Press, 2008), 2.

4. See, e.g., Ryan Chabot, "Found Innocent: Revealing the Law's Narrative Child Witnesses," *Law & Literature* 24 (Fall 2012): 319, 322.

5. Michael Burger, "Environmental Law/Environmental Literature," *Ecology Law Quarterly* 40 (2013): 2, http://dx.doi.org/doi:10.15779/Z386G42.

6. Laura King, "Narrative, Nuisance, and Climate Change," *Journal of Environmental Law and Litigation* 29 (2004): 331, 333.

7. Stephen M. Johnson, "From Climate Change and Hurricanes to Ecological Nuisances: Common Law Remedies for Public Law Failures?," *Georgia State University Law Review* 27 (2011): 565, 567.

8. John B. Wright, "Land Tenure: The Spatial Musculature of the American West," in Gary Hausladen, ed., *Western Places, American Myths: How We Think about the West* (Reno: University of Nevada Press, 2003), 85, 88.

9. Ibid.

10. Ibid.

11. John O'Sullivan, "Annexation," *United States Magazine and Democratic Review* 17 (1845): 5–10.

12. Julius W. Pratt, "The Origin of 'Manifest Destiny,'" *American Historical Review* 32 (July 1927): 795–798.

13. Shane Mountjoy, *Manifest Destiny: Westward Expansion* (New York: Chelsea House Publishing, 2009), 19.

14. Donald Worster, *Dust Bowl: The Southern Plains in the 1930s* (New York: Oxford University Press, 2004), 6.

15. Ibid.

16. David G. Horrell, Cherryl Hunt, and Christopher Southgate, "Appeals to the Bible in Ecotheology and Environmental Ethics: A Typology of Hermeneutical Stances," *Studies in Christian Ethics* 21 (2008): 219.

17. Millennium Ecosystem Assessment, *Ecosystems and Human Well-Being: Synthesis* (Washington, DC: Island Press, 2005): 1.

18. Ibid.

19. Rachel Carson, *Silent Spring* (Boston: Houghton Mifflin, 1962).

20. See generally Don Nardo, *The Blue Marble: How a Photograph Revealed Earth's Fragile Beauty* (North Mankato, MN: Compass Point Books, 2014).

21. "Dust Bowl," History.com, http://www.history.com/topics/dust-bowl# (as viewed January 16, 2016).

22. John V. Byrne, "Salmon Is King—Or Is It?," *Environmental Law* 16 (1986): 343, 346–354.

23. National Oceanic and Atmospheric Administration, NOAA Fisheries, "Atlantic Salmon *(Salmo salar)*" (as updated May 14, 2015), http://www.fisheries.noaa.gov/pr/species/fish/atlantic-salmon.html.

24. Oliver A. Houck, *The Clean Water Act TMDL Program: Law, Policy, and Implementation* (Washington, DC: Environmental Law Institute, 2002).

25. Research published in 2013 indicates that "as the average nation grows the number of endangered species increases by 3% every ten years," that "11% of animals worldwide will be endangered by 2050," and that "humans are the leading cause of animal extinction." Nicola Rowe, "Humans ARE Directly to Blame for a Rise in the Number of Endangered Species, Claims Scientists," *Daily Mail Australia* (June 22, 2013), http://www.dailymail.co.uk/sciencetech/article-2345874/Humans-ARE-directly-blame-rise-number-endangered-species-claims-scientists.html.

26. Burger, "Environmental Law/Environmental Literature," 3–4. Burger identifies four important ecological narratives recurring throughout US environmental and natural resources law: the pastoral; wilderness and wildness; the "environmental apocalyptic"; and "toxic tales."

27. Brian Walker and David Salt, *Resilience Thinking: Sustaining Ecosystems and People in a Changing World* (Washington, DC: Island Press, 2006), 6.

28. 33 U.S.C. § 1311 (2012).

29. 42 U.S.C. §§ 7411, 7479, 7501 (2012).

30. 42 U.S.C. § 300f (2012).

31. Robin Kundis Craig, "'Stationarity Is Dead'—Long Live Transformation: Five Principles for Climate Change Adaptation Law," *Harvard Environmental Law Review* 34 (2010): 32–37, http://www.law.harvard.edu/students/orgs/elr/vol34_1/9-74.pdf.," 9, 32.

32. 33 U.S.C. § 1251(a) (2012).

33. 42 U.S.C. § 9607(a)(4) (2012); 43 C.F.R. § 11.10(e)(3) (2015).

34. 33 U.S.C. §§ 2702(b)(2)(A), 2706(b)(2)(A) (2012); 33 C.F.R. § 136.211(a) (2015).

35. 42 U.S.C. § 6924(u), (v); 40 C.F.R. §§ 257.21–.28; 258.50, 258.51 (2015).

36. 30 U.S.C. § 1265(a), (b)(2) (2012).

37. 16 U.S.C. §§ 1531(b), 1532(3) (2006).

38. 16 U.S.C. §1536(g), (h) (2012).

39. Walker and Salt, *Resilience Thinking,* 6.

40. Craig, "'Stationarity Is Dead,'" 9, 32.

41. Ibid.

42. Daniel B. Botkin, *Discordant Harmonies* (1992); see also Daniel B. Botkin, "No: There Is No Balance of Nature, but We Keep Acting as If There Was" (May 23, 2013), http://www.danielbbotkin.com/2013/05/23/is-there-a-balance-of-nature/.

43. Intergovernmental Panel on Climate Change (IPCC), *Climate Change 2014: Synthesis Report* (2014), available at http://www.ipcc.ch.

44. We provide much more detail regarding these policy efforts in our law review article on this subject. See Robin Kundis Craig and Melinda Harm Benson, "Replacing Sustainability," *Akron Law Review* 46 (2013): 841, 842.

45. See Ted Nordhaus and Michael Shellenberger, *Break Through: From the Death of Environmentalism to the Politics of Possibility* (Boston: Houghton Mifflin, 2007), 160–161.

46. See "Trust in Government Nears Record Low, but Most Federal Agencies Are Viewed Favorably," Pew Center for People and the Press (accessed May 12, 2014), http://www.people-press.org/2013/10/18/trust-in-government-nears-record-low-but-most-federal-agencies-are-viewed-favorably.

47. Bill McKibben, *The End of Nature* (New York: Random House, 2006).

48. Michael Burger, Elizabeth Burleson, Rebecca M. Bratspies, Robin Kundis Craig, Alexandra R. Harrington, Keith H. Hirokawa, Sarah Krakoff, Katrina Fischer Kuh, Stephen R. Miller, Jessica Owley, Patrick A. Parenteau, Melissa Powers, Shannon Roesler, and Jonathan D. Rosenbloom, "Rethinking Sustainability to Meet the Climate Change Challenge," *Environmental Law Reporter* 43 (2012): 10342, 10355.

49. *Report of the World Commission on Environment and Development: Our Common Future,* 27, United Nations, General Assembly Resolution 42/187 (11 December 1987), http://www.un-documents.net/our-common-future.pdf.

50. See generally Jeffery D. Kovar, "A Short Guide to the Rio Declaration," *Colorado Journal of International Environmental Law & Policy* 4 (1993): 119.

51. John C. Dernbach and Federico Cheever, "Sustainable Development and Its Discontents," *Transnational Environmental Law* 4 (2015): 247–287, doi:10.1017/S2047102515000163.

52. Norman Myers, "Consumption: Challenge to Sustainable Development," *Science* 276 (1997): 53–54.

53. United Nations Environment Programme, "Global Environment Outlook—5" (2012), 21–22, http://www.unep.org/geo/geo5.asp.

54. United Nations Environment Programme, Press Release, "World Remains on Unsustainable Track Despite Hundreds of Internationally Agreed Goals and Objectives" (June 6, 2012), http://www.unep.org/geo/.

55. See Craig, "'Stationarity Is Dead,'" 9, 10–16, 23–27; P. C. D. Milly et al., "Stationarity Is Dead: Whither Water Management?," *Science* 319 (2008): 573.

56. See generally Adrian Parr, *Hijacking Sustainability* (Cambridge, MA: MIT

Press, 2009), describing the commodification of the sustainable development concept.

57. Burger et al., "Rethinking Sustainability to Meet the Climate Change Challenge," 10356.

58. W. M. Adams, *The Future of Sustainability: Re-Thinking Environment and Development in the Twenty-first Century,* IUCN, the World Conservation Union (May 2006), 1, http://cmsdata.iucn.org/downloads/iucn_future_of_sustanability.pdf.

59. Jennifer A. Elliott, *An Introduction to Sustainable Development,* 4th ed. (London: Routledge, 2013), 8.

60. World Commission on Environment and Development, "Our Common Future" (1987), 43, http://www.un-documents.net/our-common-future.pdf.

61. Elliott, *An Introduction,* 9.

62. Ibid., 10, 12–14 tbl. 1.2.

63. Ibid., 16.

64. Ibid., 8.

65. Ibid., 18 (quoting K. Lee, A. Holland, and D. McNeill, eds., *Global Sustainable Development in the Twenty-First Century* [Edinburgh: Edinburgh University Press, 2000], 9).

66. Adams, *The Future of Sustainability,* 21.

67. Ibid., 2 fig. 1.

68. Ibid, 2.

69. Elliott, *An Introduction,* 8.

70. Ibid.

71. Adams, *The Future of Sustainability,* 2.

72. Ibid., 1.

73. Ibid.

74. Ibid., 3.

75. Ibid., 31.

76. IPCC, *Climate Change 2014: Impacts, Adaptation, and Vulnerability. Part A: Global and Sectoral Aspects. Contribution of Working Group II to the Fifth Assessment Report of the Intergovernmental Panel on Climate Change* (Cambridge: Cambridge University Press, 2014), 1104, http://www.ipcc.ch.

77. Ibid.

78. Jeffrey D. Sachs, *The Age of Sustainable Development* (New York: Columbia University Press 2015), 40.

79. Adams, *The Future of Sustainability,* 11.

80. Global Footprint Network, Earth Overshoot Day 2016, as viewed July 1, 2016, http://www.overshootday.org.

81. Adams, *The Future of Sustainability,* 4–5, 6 Box 1.

82. Johan Rockström and Mattias Klum, *Big World, Small Planet: Abundance within Planetary Boundaries* (New Haven, CT: Yale University Press, 2015), 15.

83. Ibid., 60.

84. Ibid., 59.

85. Ibid., 61.

86. Ibid., 71.

87. Ibid., 65 fig. 2.1.

88. Thomas Malthus, *An Essay on the Principle of Population* (Oxford World's Classics reprint 1798), 61.

89. Paul R. Ehrlich, *The Population Bomb* (New York: Ballantine Books, 1968).

90. Linus Blomqvist, Ted Nordhaus, and Michael Shellenberger, *Nature Unbound: Decoupling for Conservation* (September 2015), http://thebreakthrough.org /images/pdfs/Nature_Unbound.pdf.

91. Ibid.

92. Rockström and Klum, *Big World, Small Planet*, 21.

93. Ibid., 39.

94. Ibid., 11–12.

95. Adams, *The Future of Sustainability*, 7.

96. Ibid., 9.

97. Ibid., 8.

98. Ibid., 13.

99. Ibid., 16.

100. Rockström and Klum, *Big World, Small Planet*, 18.

101. Ibid., 41.

102. Ibid., 18.

103. "Climate Change: The 'Greatest Threat' to the Peoples of the Pacific," *Island Business* (July 31, 2014), http://www.islandsbusiness.com/news/palau/5906/climate -change-the-greatest-threat-to-the-peoples-/.

Chapter 3. Resilience and the Trickster: A New Narrative for the Anthropocene

Parts of this chapter were developed earlier in: Robin Kundis Craig, "Putting Resilience Theory into Practice: The Example of Fisheries Management," *Natural Resources & Environment* 31, no. 3 (Winter 2017): 3–17; Robin Kundis Craig, "Learning to Live with the Trickster: Narrating Climate Change and the Value of Resilience Thinking," *Pace Environmental Law Review* 33 (Spring 2016): 351–396; Melinda Harm Benson, "Reconceptualizing Social-Ecological Relations—Is Resilience the New Narrative?," *Journal of Environmental and Sustainability Law* 21 (2015): 99–127; Melinda Harm Benson and Robin Kundis Craig, "The End of Sustainability," *Society & Natural Resources: An International Journal* 27 (2014): 777–782, DOI:10.1080/08941920.2014.9 01467; Robin Kundis Craig and Melinda Harm Benson, "Replacing Sustainability," *Akron Law Review* 46, no. 4 (2013): 841–880. Use of these previous works either conforms with the original copyright or is with permission of the publisher.

1. Tricksters are almost always male. Lewis Hyde, *Trickster Makes This World: Mischief, Myth, and Art* (New York: Farrar, Straus and Giroux, 2010 paperback edition), 8.

2. Ibid., 9.

3. Lance H. Gunderson and C. S. Holling, *Panarchy: Understanding Transformations in Human and Natural Systems* (Washington, DC: Island Press, 2002).

4. "Folklore and traditional mytho-historical narratives offer an alternative approach to framing anthropogenic and other causes of environmental change, one that has existed since the dawn of humans' capacity to historicize their lives and place in the cosmos. These narratives arguably have much to teach us about framing our understanding and contingent responses to environmental change over time and across spaces. They remind us of the futility of a managerialism that governs only for control and stability without proper consideration of relational feedbacks and the dynamic and anarchic forces in nature." Thomas F. Thornton and Patricia M. Thornton, "The Mutable, the Mythical, and the Managerial: The Mutable, the Mythical, and the Managerial," *Environment and Society* 6 (2015): 68.

5. Ibid.

6. Michael Chabon, "Foreword," in Hyde, *Trickster Makes This World*.

7. "Tricksters," *Myths Encyclopedia,* http://www.mythencyclopedia.com/Tr-Wa/Tricksters.html (as viewed January 17, 2016).

8. Hyde, *Trickster Makes This World*, 6.

9. Ibid., 7.

10. Ibid., 6–7, 9.

11. Ibid., 12.

12. Ibid., 9.

13. Ibid., 13.

14. Ibid., 6.

15. Lewis Hyde generalizes this point to argue that the trickster is largely absent from the nonpolytheistic modern world of the twentieth and twenty-first centuries, but he also argues: (1) that if modern America has a trickster, it is the confidence man, with the result that the trickster is either nowhere or everywhere in American culture; and (2) that white European settlers often became corporeal embodiments of the trickster for various Native American cultures. Ibid., 9–12.

16. Ibid., 18–19.

17. Ibid., 19.

18. Ibid., 19–23.

19. Ibid., 26.

20. Adapted from Richard Erdoes and Alfonso Ortiz, *American Indian Trickster Tales* (London: Penguin Books, 1999), 254–258.

21. Ibid, 6–9.

22. Ibid., 13–15.

23. Thornton and Thornton, "The Mutable, the Mythical, and the Managerial," 68.

24. James C. Scott, *Seeing Like a State: How Certain Schemes to Improve the Human Condition Have Failed* (New Haven, CT: Yale University Press, 1998), 6.

25. Thornton and Thornton, "The Mutable, the Mythical, and the Managerial," 66–86.

26. Ibid.

27. Ibid.

28. Masato Mori, Masahiro Watanabe, Hideo Shiogama, Jun Inoue, and Masahide Kimoto, "Robust Arctic Sea-Ice Influence on the Frequent Eurasian Cold Winters in Past Decades," *Nature Geoscience* 7 (2014): 869–873.

29. E.g., Thomas L. Friedman, "Global Weirding Is Here," *New York Times,* February 17, 2010, http://www.nytimes.com/2010/02/17/opinion/17friedman.html?_r=0.

30. Thornton and Thornton, "The Mutable, the Mythical, and the Managerial," 75.

31. As Working Group II stated in its 2014 report, "climate change calls for new approaches to sustainable development that take into account complex interactions between climate and social and ecological systems." Intergovernmental Panel on Climate Change, *Climate Change 2014: Impacts, Adaptation, and Vulnerability* (Cambridge: Cambridge University Press, 2014), 1104, http://www.ipcc.ch [hereinafter 2014 IPCC Adaptation Report].

32. C. S. Holling, "Resilience and Stability of Ecological Systems," *Annual Review of Ecology, Evolution, and Systematics* 4 (1973): 1.

33. See Brian Walker and David Salt, *Resilience Practice* (London: Island Press, 2012), 3.

34. Carl Folke, Steve Carpenter, Thomas Elmqvist, Lance Gunderson, C. S. Holling, and Brian Walker, "Resilience and Sustainable Development: Building Adaptive Capacity in a World of Transformations," *AMBIO: A Journal of the Human Environment* 31(2002): 437–440.

35. Linda Booth Sweeney, *When a Butterfly Sneezes: A Guide for Helping Kids Explore Interconnections in Our World through Favorite Stories* (Waltham, MA: Pegasus Communications, 2001), 13.

36. Ibid., 20.

37. Ibid., 17, 19.

38. John H. Miller and Scott E. Page, *Complex Adaptive Systems: An Introduction to Computational Models of Social Life* (Princeton, NJ: Princeton University Press: 2007), 3.

39. Neil Johnson, *Two's Company, Three Is Complexity* (Oxford: Oneworld Publications, 2007), 3–4.

40. Ibid., 15.

41. Ibid.

42. Ann P. Kinzig et al., "Resilience and Regime Shifts: Assessing Cascading Effects," *Ecology & Society* 11, no. 1 (2006), http://www.ecologyandsociety.org/vol11/iss1/art20/.

43. Resilience Alliance, "Adaptive Cycle," http://www.resalliance.org/adaptive-cycle (as viewed January 17, 2016).

44. Gunderson and Holling, *Panarchy,* 34.

45. Ibid.

46. Resilience Alliance, "Key Concepts: Panarchy," (as viewed January 17, 2016), http://www.resalliance.org/panarchy.

47. Ibid.

48. Ibid.

49. Kinzig et al., "Resilience and Regime Shifts: Assessing Cascading Effects," 20.

50. Ibid.

51. Ibid.

52. Walker and Salt, *Resilience Thinking,* 100.

53. Ibid., 101.

54. Joyeeta Gupta, Catrien Termeer, Judith Klostermann, Sander Meijerink, Margo van den Brink, Pieter Jong, Sibout Nooteboom, and Emmy Bergsma, "The Adaptive Capacity Wheel: a Method to Assess the Inherent Characteristics of Institutions to Enable the Adaptive Capacity of Society," *Environmental Science and Policy* 13 (2010): 459, 461.

55. Walker and Salt, *Resilience Thinking,* 7–9.

56. See Steven Heydemann and Reinoud Leenders, "Authoritarian Learning and Authoritarian Resilience: Regime Responses to the 'Arab Awakening,'" *Globalizations* 8 (2011): 647, 652 (arguing that the spread of protests throughout the Arab world can be viewed as the product of social learning by Arab citizens).

57. Stephen Carpenter, Brian Walker, J. Martin Anderies, and Nick Abel, "From Metaphor to Measurement: Resilience of What to What?," *Ecosystems* 4 (2001): 765 (identifying various definitions of resilience).

58. Stephen R. Carpenter and William A. Brock, "Adaptive Capacity and Traps," *Ecology and Society* 13 (2008): 40, http://www.ecologyandsociety.org/vol13/iss2/art40/; see generally Carl Folke, Johan Colding, and Fikret Berkes, "Building Resilience for Adaptive Capacity in Social-Ecological Systems," *Navigating Social-Ecological Systems: Building Resilience for Complexity and Change* (Cambridge: Cambridge University Press: 2002).

59. W. M. Adams, *The Future of Sustainability: Re-Thinking Environment and Development in the Twenty-first Century,* IUCN, the World Conservation Union (May 2006), 1, 12, http://cmsdata.iucn.org/downloads/iucn_future_of_sustainability.pdf.

60. 2014 IPCC Adaptation Report, 1104–1105.

61. Ibid., 1104.

62. Melinda Harm Benson, "Intelligent Tinkering: The Endangered Species Act and Resilience," *Ecology and Society* 17 (2012): 28, http://dx.doi.org/10.5751/ES-05116-170428.

63. US Fish and Wildlife Service, "Wildlife Refuges and the Next Generation," *Conserving the Future* (October 2011): 36, https://www.fws.gov/refuges/pdfs/FinalDocumentConservingTheFuture.pdf.

64. US Forest Service, *Forest Service Manual* (updated continually), https://fs.usda.gov/FSI_Directives/wo_id_2020-2011-1.doc.

65. Reed D. Benson, "New Adventures of the Old Bureau: Modern-day Reclamation Statutes and Congress's Unfinished Business," *Harvard Journal on Legislation* 48 (2011): 137, 169–172.

66. *NOAA's Next Generation Strategic Plan*, National Marine Fisheries Service, National Oceanic and Atmospheric Administration, US Department of Commerce (December 2010), v.

67. There is a spirited debate on this issue on the Resilience Science website between resilience critics and scholars on this and other issues: http://rs.resalliance.org/2009/05/19/machine-fetishism-money-and-resilience-theory/.

68. Debra Davidson, "The Applicability of the Concept of Resilience to Social Systems: Some Sources of Optimism and Nagging Doubts," *Society and Natural Resources* 23 (2005): 1135–1149.

69. 16 U.S.C. §§ 1531–1540.

70. 16 U.S.C. § 1531(b).

71. 16 U.S.C. § 1532(6), (20).

72. 16 U.S.C. § 1532(5)(A).

73. J. B. Ruhl, "Climate Change and the Endangered Species Act: Building Bridges to the No-Analog Future," *Boston University Law Review* 88 (February 2008): 1–62.

74. J. Michael Scott, Dale D. Goble, John A. Wiens, David S. Wilcove, Michael Bean, and Timothy Male, "Recovery of Imperiled Species under the Endangered Species Act: The Need for a New Approach," *Frontiers in Ecology & the Environment* 3 (2005): 383–389, doi:10.2307/3868588.

75. Millennium Ecosystem Assessment, *Ecosystems and Human Well-Being: Synthesis* (Washington, DC: Island Press, 2005).

76. 16 U.S.C. § 1601(d).

77. Benson, "Intelligent Tinkering," 28.

78. J. B. Ruhl, "General Design Principles for Resilience and Adaptive Capacity in Legal Systems: Applications to Climate Change Adaptation Law," *North Carolina Law Review* 89 (2011): 1391, http://www.nclawreview.org/documents/89/5/ruhl.pdf.

79. Ibid.

80. Ibid., 1394.

81. Ibid., 1395.

82. Ibid., 1396. Dynamic federalism is an emerging challenge to traditional notions that the division of responsibilities across scales of governance promotes optimization and efficiency. Ibid., 1397 (citing Benjamin K. Sokacool, "The Best of Both Worlds: Environmental Federalism and the Need for Federal Action on Renewable Energy and Climate Change," *Stanford Environmental Law Journal* 27 [2008]: 397, 448).

83. Ruhl, "General Design Principles," 1398.

84. Robin Kundis Craig and J. B. Ruhl, "Designing Administrative Law for Adaptive Management," *Vanderbilt Law Review* 67 (January 2014): 1–87.

85. 2014 IPCC Adaptation Report, 1104–1105.

86. Ibid.

87. 2014 IPCC Synthesis Report, 80.

88. See generally J. B. Ruhl, "The Political Economy of Climate Change Winners and Losers," *Minnesota Law Review* 97 (2012): 206; Robin Kundis Craig, "The Social and Cultural Aspects of Climate Change Winners," *Minnesota Law Review* 97 (2013): 1416 (both discussing a variety of issues arising from the fact that, at least for a while, some people will benefit from climate change).

Chapter 4. Regime Change for New Mexico Watersheds

Parts of this chapter were developed earlier in: Melinda Harm Benson, "Shifting Public Land Management Paradigms: Lessons from the Valles Caldera National Preserve," *Virginia Environmental Law Journal* 34 (2016): 1–51; Melinda Harm Benson, Dagmar Llewellyn, Ryan Morrison, and Mark Stone, "Water Governance Challenges in New Mexico's Rio Grande Valley: A Resilience Assessment," *Idaho Law Review* 51 (2014): 195–228. Use of these previous works either conforms with the original copyright or is with permission of the publisher.

1. See generally Fred M. Phillips, G. Emlen Hall, and Mary E. Black, *Reining in the Rio Grande: People, Land and Water* (Albuquerque: University of New Mexico Press, 2011); Bill Debuys, *A Great Aridness: Climate Change and the Future of the American Southwest* (New York: Oxford University Press, 2011), 63–71.

2. Phillips, Hall, and Black, *Reining in the Rio Grande,* 37–65 (providing a detailed account of early Spanish influence).

3. Ibid.

4. Utton Transboundary Resources Center, University of New Mexico (2014), "Water Matters!," http://uttoncenter.unm.edu/pdfs/water-matters-2014/2014-water-matters-lr.pdf (citing Treaty of Guadalupe Hidalgo of 1848, art III, U.S.-Mex., Feb. 2, 1848, 9 Stat. 922).

5. Ibid., 12.

6. See generally Donald J. Pisani, *Water and American Government: The Reclamation Bureau, National Water Policy in the West 1902–1935* (Berkeley: University of California Press, 2002); see also K. Maria D. Lane, "Water, Technology, and the Courtroom: Negotiating Reclamation Policy in Territorial New Mexico," *Journal of Historical Geography* 37 (2011): 300.

7. Thomas R. Karl, Jerry M. Melillo, and Thomas C. Peterson, eds., *Global Climate Change Impacts in the United States* (Cambridge: Cambridge University Press, 2009), www.globalchange.gov/usimpacts; Dagmar Llewellyn and Seshu Vaddey, U.S. Bureau of Reclamation, "West-Wide Climate Risk Assessment: Upper Rio Grande Impact Assessment 2013," http://www.usbr.gov/WaterSMART /wcra/docs/urgia/URGIAMainReport.pdf; US Bureau of Reclamation, "West-Wide Climate Risk Assessment: Upper Rio Grande Impact Assessment 2013," http://www.usbr.gov/WaterSMART/wcra/docs/urgia/URGIAMainReport.pdf.

8. Melinda Harm Benson, Dagmar Llewellyn, Ryan Morrison, and Mark Stone, "Water Governance Challenges in New Mexico's Rio Grande Valley: A Resilience Assessment," *Idaho Law Review* 51 (2014): 195–228.

9. Jeanine M. Rhemtulla, David J. Mladenoff, and Murray K. Clayton, "Regional Land-cover Conversion in the U.S. Upper Midwest: Magnitude of Change and Limited Recovery (1850–1935–1993)," *Landscape Ecology* 22 (2007): 57–75, doi:10.1007/s10980-007-9117-3.

10. Brian Walker and David Salt, *Resilience Practice* (London: Island Press, 2012).

11. Ibid.

12. Ibid, 101.

13. B. C. Chaffin, A. S. Garmestani, L. H. Gunderson, M. H. Benson, D. G. Angeler, C. A. Arnold, B. Cosens, R. K. Craig, J. B. Ruhl, and C. R. Allen, "Transformative Environmental Governance," *Annual Review of Environment and Resources* 41 (2016): 1, http://www.annualreviews.org/doi/abs/10.1146/annurev-environ-110615 -085817.

14. Stephen Carpenter, Brian Walker, J. Martin Anderies, and Nick Abel, "From Metaphor to Measurement: Resilience of What to What?," *Ecosystems* 4 (2001): 765.

15. See Stephen R. Carpenter and William A. Brock, "Adaptive Capacity and Traps," *Ecology & Society* 13 (2008): 40, http://www.ecologyandsociety.org/vol13 /iss2/art40/.

16. Renee A. O'Brien, "New Mexico's Forests 2000," US Department of Agriculture, Forest Service, *Rocky Mountain Research Station Resource Bulletin*, RMRS-RB-3, December 2003, http://www.fs.fed.us/rm/ogden/pdfs/rmrs_rb003.pdf.

17. 16 U.S.C. §§ 528–531 (2012).

18. 16 U.S.C. §§ 1600–1614 (2012).

19. Utton Transboundary Resources Center, "Water Matters!"

20. A. Park Williams, Craig D. Allen, Alison K. Macalady, Daniel Griffin, Connie A. Woodhouse, David M. Meko, Thomas W. Swetnam, Sara A. Rauscher, Richard Seager, Henri D. Grissino-Mayer, Jeffrey S. Dean, Edward R. Cook, Chandana Gangodagamage, Michael Cai, and Nate G. McDowell, "Temperature as a Potent Driver of Regional Forest Drought Stress and Tree Mortality," *Nature Climate Change* 3 (2013): 292–297, doi:10.1038/nclimate1693.

21. Ibid.

22. Ibid.

23. See USDA, PR-R3-16-8, "Forest Insect and Disease Conditions in the Southwestern Region, 2011," (2012): 3, http://www.fs.usda.gov/Internet/FSE_DOCU MENTS/stelprdb5406441.pdf.

24. Craig D. Allen, "Interactions across Spatial Scales among Forest Dieback, Fire, and Erosion in Northern New Mexico Landscapes," *Ecosystems* 10 (2007): 798, https://www.fort.usgs.gov/products/sb/6597.

25. See Jeffrey A. Hicke et al., "Effects of Bark Beetle–Caused Tree Mortality on Wildfire," *Forest Ecology & Management* 271 (2012): 81, 84, doi:0.1016/j .foreco.2012.02.005.

26. Allen, "Interactions across Spatial Scales," 801.

27. Tania Schoennagel et al., "The Interaction of Fire, Fuels, and Climate across Rocky Mountain Forests," *BioScience* 54 (2004): 661, 666, http://spot.colo rado.edu/~schoenna/images/Schoennagel2004BioScience.pdf.

28. Max A. Moritz et al., "Climate Change and Disruptions to Global Fire Activity," *Ecosphere* 3 (2012): 1, 18, http://dx.doi.org/10.1890/ES11-00345.1.

29. Allen, "Interactions across Spatial Scales," 798.

30. Sandra Postel, "Wildfires in the Western U.S. Are on the Rise, Posing Threats to Drinking Water," *National Geographic Newswatch* (April 29, 2014), http://newswatch.nationalgeographic.com/2014/04/29/wildfires-in-the-western -u-s-are-on-the-rise-posing-threats-to-drinking-water/ (last visited May 2, 2014).

31. Schoennagel et al., "The Interaction of Fire, Fuels, and Climate across Rocky Mountain Forests," 661, 666.

32. Daniel G. Neary, "Post-Wildfire Watershed Flood Responses," audio, 2nd International Wildland Fire Ecology and Fire Management Congress, November 17, 2003, https://ams.confex.com/ams/FIRE2003/techprogram/paper_65982 .htm.

33. J. Funk, S. Saunders, T. Sanford, T. Easley, and A. Markham, "Rocky Mountain Forests at Risk: Confronting Climate-Driven Impacts from Insects, Wildfires, Heat, and Drought," *Report from the Union of Concerned Scientists and the Rocky Mountain Climate Organization* (Cambridge, MA: Union of Concerned Scientists, 2014), http://www.ucsusa.org/global_warming/science_and_impacts /impacts/climate-change-impacts-rocky-mountain-forests.html#.V96oyZMrIyk.

34. Tiffany Stecker, "Carbon Capture: Fires, Urbanization to Redraw Carbon

Map of West," Environment and Energy Publishing (December 7, 2012), http://www.eenews.net/stories/1059973492.

35. Michael G. Ryan, Mark E. Harmon, Richard A. Birdsey, Christian P. Giardina, Linda S. Heath, Richard A. Houghton, Robert B. Jackson, Duncan C. McKinley, James F. Morrison, Brian C. Murray, Diane E. Pataki, and Kenneth E. Skog, "A Synthesis of the Science on Forests and Carbon for U.S. Forests," *U.S. Forest Service Issues in Ecology Report 2010,* http://www.fs.fed.us/rm/pubs_other/rmrs_2010_ryan_m002.pdf.

36. Llewellyn and Vaddey, "West-Wide Climate Risk Assessment."

37. Laura Paskus, "A Hotspot for Warming, NM Offers Lesson for Forecasters," *New Mexico in Depth* (2015), http://nmindepth.com/2015/10/27/a-hotspot-for-warming-new-mexico-offers-lesson-for-forecasters/).

38. Ibid.

39. National Forest Management Act of 1976 16 U.S.C §§ 1600–1687, and the Multiple-Use Sustained-Yield Act of 1960, 16 U.S.C. §§ 528–531.

40. Dyan Zaslowsky and T. H. Watkins, *These American Lands: Parks, Wilderness, and the Public Lands: Revised and Expanded Edition* (Washington, DC: Island Press, 1994).

41. US Forest Service, Forest Service Manual, "1900—Planning Chapter 1920 Land Management Planning" (effective date January 30, 2015), http://www.fs.usda.gov/detail/planningrule/home/?cid=stelprd3828310.

42. "21—Developing, Revising, Amending, or Administratively Changing a Plan," U.S. Forest Service Handbook, http://www.fs.usda.gov/Internet/FSE_DOCUMENTS/stelprdb5409939.pdf.

43. US Forest Service, "Forest Plan Revision Public Participation Timeline," http://www.fs.usda.gov/Internet/FSE_MEDIA/fseprd497695.jpg.

44. "Increasingly severe fire seasons are one of the greatest challenges facing the Nation's forests," Tidwell said in a prepared statement. "The cost of fire suppression has soared in the past 20 years." Elizabeth Harball, "WILDFIRE: Severe Fire Season Poised to Burn through Forest Service's Budget," June 10, 2014, http://www.eenews.net/climatewire/stories/1060001005.

45. "2013 Fire Transfer Activity; Deferring Other Financial Obligations," http://www.eenews.net/assets/2013/08/21/document_pm_01.pdf.

46. New Mexico Annual Bulletin—2014, USDA, National Agricultural Statistics Service in cooperation with New Mexico Department of Agriculture, https://www.nass.usda.gov/Statistics_by_State/New_Mexico/Publications/Annual_Statistical_Bulletin/2014/2014_NM_Pub.pdf.

47. N.M. Const. art. XVI, § 2: "The unappropriated water of every natural stream, perennial or torrential, within the state of New Mexico, is hereby declared to belong to the public and to be subject to appropriation for beneficial use, in accordance with the laws of the state. Priority of appropriation shall give the better right."

48. Ibid.

49. Utton Transboundary Resources Center, "Water Matters!"

50. Ibid.

51. Ibid.

52. S. S. Papadopulos & Assocs., "Evaluation of Middle Rio Grande Conservancy District Irrigation System and Measurement Program," ES-2 (2002), http://www.ose.state.nm.us/Pub/MRGCD/volume-1-rpt.pdf.

53. Douglas W. Strech and Tracy Scharp Matthews, "Middle Rio Grande Vegetation Classification Summer 2000," August 15, 2001 (on file with author; this was a joint project between the Middle Rio Grande Conservancy District and the New Mexico Office of the State Engineer/Interstate Stream Commission).

54. S. S. Papadopulos & Assocs., "Evaluation of Middle Rio Grande Conservancy District Irrigation System and Measurement Program."

55. Sarah Bates, "Bridging the Governance Gap: Emerging Strategies to Integrate Water and Land Use Planning," *Natural Resources Journal* 52 (2012): 61, discussing the need for integrated land use and water planning. In theory, this should not be true in the fully allocated system of the Rio Grande Watershed, where new uses need to be offset by the retirement of existing uses, primarily agriculture. In the absence of adjudication, however, accommodating other municipal water use remains problematic.

56. Utton Transboundary Resources Center, "Water Matters!"

57. Ibid.

58. "Acre-feet" is the volume of water necessary to cover one acre of surface area to a depth of one foot.

59. Utton Transboundary Resources Center, "Water Matters!"

60. US Bureau of Reclamation, "SECURE Water Act § 9503(c), Reclamation: Climate Change and Water" (2011), http://www.usbr.gov/climate/SECURE/docs/SECUREWaterReport.pdf. See also Reed D. Benson, "New Adventures of the Old Bureau: Modern-Day Reclamation Statutes and Congress' Unfinished Environmental Business," *Harvard Journal on Legislation* 48 (2011): 137, 169–72, http://papers.ssrn.com/sol3/papers.cfm?abstract_id=1621758.

61. Walker and Salt, *Resilience Practice*, 101.

62. The Landscape Conservation Cooperatives are collaborative, intergovernmental programs coordinated by the Fish and Wildlife Service, with assistance from other agencies, including Reclamation in the Desert and Southern Rockies landscapes covering the Rio Grande Watershed. See US Fish & Wildlife Service, "Landscape Conservation Cooperatives: Frequently Asked Questions" (February 2012), 1, http://www.fws.gov/landscape-conservation/pdf/LCC_FAQs_2012.pdf (last visited Jan. 17, 2015).

63. Ellis Q. Margolis and Jeff Balmat, "Fire History and Fire–Climate Relationships Along a Fire Regime Gradient in the Santa Fe Municipal Watershed, NM,"

Forest Ecology and Management 258 (2009): 2416, 2417; see also City of Santa Fe, "Municipal Watershed Plan," http://www.santafenm.gov/municipal_watershed _plan (last visited January 17, 2015).

64. City of Santa Fe, "Municipal Watershed Investment Plan," http://www .santafenm.gov/municipal_watershed_investment_plan.

65. Ibid.

66. Ibid.

67. US Department of Agriculture, Forest Service, "Proposed Action for Southwest Jemez Mountains Restoration," (2012), 1, http://www.fs.usda.gov/Inter net/FSE_DOCUMENTS/stelprdb5379537.pdf.

68. Scott Streater, "Rio Grande: Conservation Group Backs N.M. Forest Thinning to Curb Wildfires," *Greenwire,* Environment & Energy Publishing, July 29, 2014, http://www.eenews.net/greenwire/stories/1060003703/search?keyword=rio +grande+fund.

69. Carl Folke, Thomas Hahn, Per Olsson, and Jon Norberg, "Adaptive Governance of Social-Ecological Systems," *Annual Review of Environment and Resources* 30 (2005): 441–473, doi:10.1146/annurev.energy.30.050504.144511.

70. Streater, "Rio Grande: Conservation Group Backs N.M. Forest Thinning to Curb Wildfires."

71. "Adaptation: N.M. Community Tries to Build Resilience to Flood, Fire Risks," *ClimateWire,* Environment & Energy Publishing, July 28, 2015, http:// www.eenews.net/climatewire/2015/07/28/stories/1060022486.

72. Ibid.

73. Ibid.

74. Ibid.

75. N. J. Mantua, "The Pacific Decadal Oscillation and Climate Forecasting for North America," *Climate Risk Solutions* 1 (1999): 10–13.

76. Craig Allen, personal communication (November 14, 2015).

77. Max A. Moritz et al. "Learning to Coexist with Wildfire," *Nature* 515 (2014): 58–66, http://dx.doi.org/10.1038/nature13946.

78. Greg Harmon, "Your Brain on Climate Change: Why the Threat Produces Apathy, Not Action," *The Guardian,* November 10, 2014, http://www.theguard ian.com/sustainable-business/2014/nov/10/brain-climate-change-science-psychol ogy-environment-elections.

79. Robert B. Keiter, "The Law of Fire: Reshaping Public Land Policy in an Era of Ecology and Litigation," *Environmental Law* 36 (2006): 301, 313–314, http:// ssrn.com/abstract=930116.

80. Byron K. Williams, Robert C. Szaro, and Carl D. Shapiro, *Adaptive Management: The US Department of the Interior Technical Guide* (2009): 6.

81. See Keiter, "The Law of Fire: Reshaping Public Land Policy in an Era of Ecology and Litigation," 301, 313–314.

82. US Bureau of Reclamation, "SECURE Water Act § 9503(c), Reclamation: Climate Change and Water," 17–40.

83. Joyeeta Gupta, Catrien Termeer, Judith Klostermann, Sander Meijerink, Margo van den Brink, Pieter Jong, Sibout Nooteboom, and Emmy Bergsma, "The Adaptive Capacity Wheel: A Method to Assess the Inherent Characteristics of Institutions to Enable the Adaptive Capacity of Society," *Environmental Science and Policy* 13 (2010): 459, 461.

Chapter 5. Marine Fisheries and Biodiversity: How the Trickster Undermines Sustainable Yield

Parts of this chapter were explored in earlier publications, including: Robin Kundis Craig, "Putting Resilience Theory into Practice: The Example of Fisheries Management," *Natural Resources & Environment* 31, no. 3 (Winter 2017): 3–7; Robin Kundis Craig, "Dealing with Ocean Acidification: The Problem, the Clean Water Act, and State and Regional Approaches," *University of Washington Law Review* 90 (Winter 2015): 1583–1657; Robin Kundis Craig, "Re-Tooling Marine Food Supply Resilience in a Climate Change Era: Some Needed Reforms," *Seattle University Law Review* 38 (2015): 1189–1235; Robin Kundis Craig, "Climate Change, Oceans, Public Health, and the Law," *Climate Law* 4 (2014): 85–93; Robin Kundis Craig, "Ocean Governance for the 21st Century: Making Marine Zoning Climate Change Adaptable." *Harvard Environmental Law Review* 36 (June 2012): 305–350; Robin Kundis Craig, "Marine Biodiversity, Climate Change, and Governance of the Oceans," *Diversity* 4 (2012): 224–238; Robin Kundis Craig and J. B. Ruhl, "Governing for Sustainable Coasts: Complexity, Climate Change, and Coastal Ecosystem Protection," *Sustainability* 2 (2010): 1361–1388. Use of these previous works either conforms with the original copyright or is with permission of the publisher.

1. Craig Welch, "Heat Wave," *National Geographic* 230 (September 2016): 54, 61–62.

2. Ibid., 60–61.

3. Ibid.

4. Ibid., 62–63.

5. Ibid., 61.

6. Ibid., 63.

7. Marine Stewardship Council, "Fish as Food," accessed February 12, 2015, http://www.msc.org/healthy-oceans/the-oceans-today/fish-as-food.

8. Ibid.

9. Ibid.

10. Karen Weintraub, "Giant Coral Reef in Protected Area Shows New Signs of Life," *New York Times,* August 15, 2016, http://www.nytimes.com/2016/08/16/science/coral-reef-climate-change.html?emc=eta1&_r=0.

11. Ibid.

12. Ibid.

13. For discussions of the benefits of marine protected areas for marine ecosystems like coral reefs, see Robin Kundis Craig, "Taking the Long View of Ocean Ecosystems: Historical Science, Marine Restoration, and the Oceans Act of 2000," *Ecology Law Quarterly* 29 (2002): 649; Robin Kundis Craig, "Protecting International Marine Biodiversity: International Treaties and National Systems of Marine Protected Areas," *Journal of Land Use & Environmental Law* 20 (2005): 333; Robin Kundis Craig, *Comparative Ocean Governance: Place-Based Protections in an Era of Climate Change* (Cheltenham, UK: Edward Elgar, 2012).

14. Sylvia Rowley, "How Dwindling Fish Stocks Got a Reprieve," *New York Times,* April 19, 2016, http://opinator.blogs.nytimes.com/2016/04/19/how-dwindling-fish-stocks-got-a-reprieve/?emc=eta1&_r=0.

15. Millennium Ecosystem Assessment, *Ecosystems and Human Well-Being: Current State and Trends* (Washington, DC: Island Press, 2005), 497 [hereinafter 2005 MEA Current State and Trends Report].

16. National Aeronautic & Space Administration (NASA), "NASA Oceanography" (last visited August 3, 2010), http://science.nasa.gov/earth-science/oceano graphy/.

17. 2005 MEA Current State and Trends Report, 480, 488.

18. Helen Scales, "National Geographic News: Tagged Tuna Reveal Migration Secrets," *National Geographic* (August 13, 2007), http://news.nationalgeographic .com/news/2007/08/070813-tuna-tags.html.

19. Robert Costanza et al., "The Value of the World's Ecosystem Services and Natural Capital," *Nature* 387 (1997): 253, 259, http://www.esd.ornl.gov/benefits _conference/nature_paper.pdf.

20. Ibid., 256.

21. Millennium Ecosystem Assessment, 2005, *Ecosystems and Human Well-being: Synthesis* 6 (Washington, DC: Island Press, 2005) [hereinafter 2005 MEA Synthesis Report].

22. Ibid.

23. United Nations Education, Scientific, and Cultural Organization (UNESCO), "Marine Biodiversity," *Connect: International Science, Technology, & Environmental Education Newsletter* 21 (March 1996): 1.

24. G. Carleton Ray and J. Frederick Grassle, "Marine Biological Diversity: A Scientific Program to Help Conserve Marine Biological Diversity Is Urgently Required," *Bioscience* 41 (1991): 453, https://marine.rutgers.edu/pubs/private/grassle /1991%20Marine%20biological%20diversity.pdf.

25. UNESCO, "Marine Biodiversity," 3.

26. Ibid.

27. Ibid., 6, 31.

28. Ibid., 31.

29. 2005 MEA Current State and Trends Report, 515.

30. Ibid., 516.

31. United Nations Environment Programme, "What Is Marine Pollution and How Does It Affect Marine Life," accessed July 28, 2010, http://www.grida.no /publications/rr/our-precious-coasts/page/1292.aspx and fig. 8.

32. Ibid.

33. Ibid.

34. Ibid. (citations omitted).

35. Nicholas Bax, Angela Williamson, Max Aguero, Exequiel Gonzalez, and Warren Geeves, "Marine Alien Species: A Threat to Global Biodiversity," *Marine Policy* 27 (2003): 313, 314.

36. Ibid., 313–314, 315.

37. Ibid., 317.

38. Ibid., 313–23, 313, 314.

39. Ibid., 313.

40. Ibid., 317.

41. Ibid., 314.

42. Robert J. Diaz and Rutger Rosenberg, "Spreading Dead Zones and Consequences for Marine Ecosystems," *Science* 321 (2008): 926, 927.

43. Bax, Williamson, Aguero, Gonzalez, and Geeves, "Marine Alien Species," 313, 314.

44. Ibid.

45. Jennifer Vargas, "Gulf Wildlife 'Dead Zone' Keeps Growing," *Discovery News* (May 7, 2010), http://news.discovery.com/animals/gulf-dead-zone-oil-spill .html.

46. See Diaz and Rosenberg, "Spreading Dead Zones," 926 ("Dead zones have developed in continental seas, such as the Baltic, Kattegat, Black Sea, Gulf of Mexico, and East China Sea, all of which are major fishery areas.").

47. Ibid., 926, 928.

48. James Owen, "World's Largest Dead Zone Suffocating Sea," *National Geographic News* (March 5, 2010), http://news.nationalgeographic.com/news/2010 /02/100305-baltic-sea-algae-dead-zones-water/.

49. See Diaz and Rosenberg, "Spreading Dead Zones," 927.

50. State of California, "Facts about Marine Pollution," Department of Boating and Waterways, 2 (2007), http://www.dbw.ca.gov/Pubs/Pollute/MarinePollution .pdf.

51. Richard Grant, "Drowning in Plastic: The Great Pacific Garbage Patch Is Twice the Size of France," *Telegraph* (U.K.), April 24, 2009, http://www.telegraph .co.uk/earth/environment/5208645/Drowning-in-plastic-The-Great-Pacific -Garbage-Patch-is-twice-the-size-of-France.html; Jacob Silverman, "Why Is the

World's Biggest Landfill in the Pacific Ocean?," (last viewed August 9, 2010), 2, http://science.howstuffworks.com/environmental/earth/oceanography/great-pacific-garbage-patch1.htm.

52. Grant, "Drowning in Plastic."

53. Silverman, "Why Is the World's Biggest Landfill in the Pacific Ocean?"

54. Grant, "Drowning in Plastic."

55. Silverman, "Why Is the World's Biggest Landfill in the Pacific Ocean?"

56. Katherine Butler, "New Ocean Garbage Patch Discovered," *Mother Nature Network,* August 4, 2010, http://www.mnn.com/earth-matters/wilderness-resources/stories/new-ocean-garbage-patch-discovered.

57. D. Malakoff, "Trash Fish: Fish in Pacific 'Garbage Patch' Ingesting Plastic," Conservation Magazine Online, June 30, 2011, http://www.conservationmagazine.org/2011/07/trash-fish/ (accessed on 14 March 2012).

58. 2005 MEA Current State and Trends Report, 483.

59. Natural Resources Defense Council, "Mercury Contamination in Fish: Know Where It's Coming From," http://www.nrdc.org/health/effects/mercury/sources.asp (last visited December 2, 2010).

60. Ibid.

61. Ibid.

62. US Environmental Protection Agency, "2011 National Listing of Fish Advisories," National Fish & Wildlife Contamination Program, (2013), 2, https://www.epa.gov/sites/production/files/2015-06/documents/maps-and-graphics-2011.pdf; US Environmental Protection Agency, 2011 National Listing of Fish Advisories: Technical Fact Sheet (December 2013), 3 fig. 2.

63. US Department of State, "Marine Biodiversity" (as viewed August 15, 2017), http://www.state.gov/e/oes/ocns/opa/biodiversity.

64. See, e.g., Owen, "World's Largest Dead Zone Suffocating Sea" (noting that overfishing in the Baltic Sea has intensified the problems caused by the dead zones there).

65. Jeremy B. C. Jackson, Michael X. Kirby, Wolfgang H. Berger, Karen A. Bjorndal, Louis W. Botsford, Bruce J. Bourque, Roger H. Bradbury, Richard Cooke, Jon Erlandson, James A. Estes, Terence P. Hughes, Susan Kidwell, Carina B. Lange, Hunter S. Lenihan, John M. Pandolfi, Charles H. Peterson, Robert S. Steneck, Mia J. Tegner, and Robert R. Warner, "Historical Overfishing and the Recent Collapse of Coastal Ecosystems," *Science* 293 (2001): 629.

66. Ibid.

67. Ibid., 634.

68. Ibid.

69. B. Wright, "Predators Could Help Save Pollock," *Science* 327 (2010): 642.

70. Daniel Pauly, Villy Christensen, Johanne Dalsgaard, Rainer Froese, and Francisco Torres Jr., "Fishing Down Marine Food Webs," *Science* 279 (1998): 860–863.

71. Joel M. Levy, ed., "Global Oceans," in State of the Climate in 2009, *Bulletin of the American Meteorological Association* 91, no. 7 (July 2010): S59.

72. Ibid., S53; see also ibid., S58 fig. 3.7 (graphing upward trend of ocean heat content since 1994).

73. 2014 IPCC Synthesis Report, 40.

74. Levy, "Global Oceans," S53–55.

75. Ibid., S55 fig. 3.3.

76. 2014 IPCC Synthesis Report, 40.

77. Tim P. Barnett, David W. Pierce, and Reiner Schnur, "Detection of Anthropogenic Climate Change in the World's Oceans," *Science* 292 (2001): 270, 271 and fig. 2 (reporting detection of increases in some oceans' temperatures to depths of at least 3,000 meters, when there are 1609.344 meters in a mile).

78. 2014 IPCC Synthesis Report, 60, 61 fig. 2.2.

79. UN Food and Agriculture Organization, "The State of World Fisheries and Aquaculture 2008" (2009): 87, ftp://ftp.fao.org/docrep/fao/011/i0250e/i0250e.pdf.

80. Shelley DaWicki, "North Atlantic Fish Populations Shifting as Ocean Temperatures Warm," Northeast Fisheries Science Center, NOAA (November 2, 2009), http://www.nefsc.noaa.gov/press_release/2009/SciSpot/SS0916/. See also B. Planque and T. Frédou, "Temperature and the Recruitment of Atlantic Cod (Gadus morhua)," *Canadian Journal of Fisheries & Aquatic Science* 56 (1999): 2069 (reporting similar results for cod).

81. See Julie M. Roessig et al., "Effects of Global Climate Change on Marine and Estuarine Fishes and Fisheries," *Reviews in Fish Biology & Fisheries* 14 (2004): 262–263 (explaining the limited options for polar fish species). According to the MEA, "climate change, acting through changes in sea temperature and especially wind patterns, will disturb and displace fisheries. Disruptions in current flow patterns in marine and estuarine systems, including changes to freshwater inputs as predicted under climate change, may cause great variations in reproductive success." 2005 MEA Current State and Trends Report, 498.

82. The MEA indicated that marine extinctions resulting directly from climate change will probably be rare, although local extirpations are likely. 2005 MEA Current State and Trends Report, 490.

83. Ibid., 489.

84. Roessig et al., "Effects of Global Climate Change."

85. Derek P. Tittensor et al., "Global Patterns and Predictors of Marine Biodiversity across Taxa," *Nature* 466 (2010): 1098.

86. 2005 MEA Current State and Trends Report, 490.

87. Woods Hole Oceanographic Institution, "Ocean Conveyor's 'Pump' Switches Back On," http://www.whoi.edu/page.do?pid= 12455&tid=282&cid=54347 (last updated November 19, 2010).

88. Ibid.

89. Ibid.

90. 2014 IPCC Synthesis Report, 60, 62.

91. F. Chan et al., "Emergence of Anoxia in the California Current Large Marine Ecosystem," *Science* 319 (2008): 920, http://www.sciencemag.org/content/319/5865/.

92. "Oregon Dead Zone Blamed on Climate Change," *Environmental News Service*, October 8, 2009, http://ens-newswire.com/2009/10/08/oregon-dead-zone-blamed-on-climate-change/.

93. Ibid.

94. Ibid.

95. Anny Cazenave, "How Fast Are the Ice Sheets Melting?," *Science* 314 (2006): 1250. But see also Mark F. Meier et al., "Glaciers Dominate Eustatic Sea-Level Rise in the 21st Century," *Science* 317 (2007): 1064, 1065 (arguing that glaciers and ice caps "contribute about 60% of the eustatic, new-water component of sea-level rise").

96. 2014 IPCC Synthesis Report, 40.

97. Cazenave, "How Fast Are the Ice Sheets Melting?," 1250; J. L. Chen et al., "Satellite Gravity Measurements Confirm Accelerated Melting of Greenland Ice Sheet," *Science* 313 (2006): 1958.

98. Cazenave, "How Fast Are the Ice Sheets Melting?," 1251 ("The greatest uncertainty in sea-level projections is the future behavior of the ice sheets").

99. 2014 IPCC Synthesis Report, 62.

100. Vivian Gornitz, "Sea Level Rise, after the Ice Melted and Today," Goddard Institute for Space Studies, NASA (January 2007), http://www.giss.nasa.gov/research/briefs/gornitz_09/; Antarctic Treaty Consultative Meeting XXIX, The Antarctic and Climate Change Report 3 (2006), http://www.asoc.org/storage/documents/Meetings/ATCM/XXIX/climate%20change%20ip.pdf; Cazenave, "How Fast Are the Ice Sheets Melting?," 1250.

101. 2014 IPCC Synthesis Report, 62.

102. Ana Swanson, "Nearly 40 Percent of Americans Live Near the Coast," *Washington Post* (September 9, 2015), http://knowmore.washingtonpost.com/2015/09/09/nearly-40-percent-of-americans-live-near-the-coast/.

103. Peter M. Cox, Richard A. Betts, Chris D. Jones, Steven A. Spall, and Ian J. Totterdell, "Acceleration of Global Warming Due to Carbon-Cycle Feedbacks in a Coupled Climate Model," *Nature* 408 (November 9, 2000): 184.

104. "The Ocean Carbon Cycle," *Harvard Magazine* (November–December 2002), http://harvardmagazine.com/2002/11/the-ocean-carbon-cycle.html.

105. Richard A. Feely, Christopher L. Sabine, and Victoria J. Fabry, "Carbon Dioxide and Our Ocean Legacy," NOAA (April 2006): 1, available at http://www.pmel.noaa.gov/pubs/PDF/feel2899/feel2899.pdf.

106. As a general matter, the cold water at ocean depths can sequester more

CO$_2$ than warmer waters at the surface (see "The Ocean Carbon Cycle"). As a result, any process that circulates cold water to the surface will reduce an ocean's ability to act as a carbon sink. Research published in 2009 indicated that, as a result of climate change, the Southern Indian Ocean is being subjected to stronger winds. The winds, in turn, mix the ocean waters, bringing up CO$_2$ from the depths and preventing the ocean from absorbing more carbon dioxide from the atmosphere. CNRS (Le Centre national de la recherche scientifique, Délégation Paris Michel-Ange), "Ocean Less Effective at Absorbing Carbon Dioxide Emitted by Human Activity," *Science Daily* (February 23, 2009), http://www.sciencedaily .com /releases/2009/02/090216092937.htm. For similar reasons, "the CO$_2$ sink diminished by 50% between 1996 and 2005 in the North Atlantic." Ibid.

107. "Ocean Acidification: Another Undesired Side Effect of Fossil Fuel-burning," *Science Daily* (May 24, 2008), http://www.sciencedaily.com/releases/2008 /05/080521105151.htm.

108. Intergovernmental Panel on Climate Change, "Climate Change 2007: Synthesis Report" (2007), 52 [hereinafter 2007 IPCC Synthesis Report].

109. Feely, Sabine, and Fabry, "Carbon Dioxide and Our Ocean Legacy," 2.

110. 2007 IPCC Synthesis Report, 52.

111. Feely, Sabine, and Fabry, "Carbon Dioxide and Our Ocean Legacy," 2.

112. Fred Pearce, *With Speed and Violence: Why Scientists Fear Tipping Points in Climate Change* (Boston: Beacon Press, 2008): 87–88.

113. Ibid., 88.

114. Peter D. Ward, *Under a Green Sky: Global Warming, the Mass Extinctions of the Past and What They Can Tell Us about Our Future* (New York: Harper Perennial, 2008), 121.

115. Joan A. Kleypas and Kimberly K. Yates, "Coral Reefs and Ocean Acidification," *Oceanography* 22 (December 2009): 108, 109.

116. Ibid. See also Sarah R. Cooley, Hauke L. Kite-Powell, and Scott C. Doney, "Ocean Acidification's Potential to Alter Global Marine Ecosystem Services," *Oceanography* 22 (December 2009): 172, 172–176 (detailing these ecosystem impacts).

117. Scott C. Doney et al., "Ocean Acidification: A Critical Emerging Problem for the Ocean Sciences," *Oceanography* 22 (2009): 16, 24.

118. 2014 IPCC Synthesis Report, 67.

119. Shawn Booth and Dirk Zeller, "Mercury, Food Webs, and Marine Mammals: Implications of Diet and Climate Change for Human Health," *Environmental Health Perspectives* 113 (2005): 521, 525.

120. Jessica R. Ward and Kevin D. Lafferty, "The Elusive Baseline of Marine Disease: Are Diseases in Ocean Ecosystems Increasing?," *PLoS Biology* 2 (2004): 542, 542–543.

121. United Nations Environment Programme World Conservation Monitoring Center, "Climate Change and Marine Diseases: The Socio-Economic Impact"

(2009), 1, http://www.unep-wcmc.org/marine/pdf/Epublication_V3_23092009
.pdf (current version as of Sept. 30, 2010).

122. 2005 MEA Current State and Trends Report, 484; Levy, "Global Oceans," S75.

123. Levy, "Global Oceans," S75.

124. Ibid., S53. See also ibid., S78 and fig. 3.33 ("From 1999 onward, an overall progressive decrease in chlorophyll is observed and coincident with a general increasing trend in surface-ocean temperature").

125. Ibid., S77–S78.

126. UN Food and Agriculture Organization, "The State of World Fisheries and Aquaculture 2008" (2009), 87.

127. International Programme on the State of the Ocean, State of the Ocean Report 4 (2013), http://www.stateoftheocean.org/wp-content/uploads/2015/10/State-of-the-Ocean-2013-report.pdf.

128. William W. L. Cheung, Reg Watson, and Daniel Pauly, "Signature of Ocean Warming in Global Fisheries Catch," *Nature* 497 (16 May 2013): 365–369, doi:10.1038/nature12156. According to NOAA, "Large Marine Ecosystems (LMEs) are relatively large areas of ocean space of approximately 200,000 km² or greater, adjacent to the continents in coastal waters where primary productivity is generally higher than in open ocean areas." NOAA, "Large Marine Ecosystems of the World," (as viewed August 9, 2016), http://www.lme.noaa.gov.

129. Cheung, Watson, and Pauly, "Signature of Ocean Warming," 368.

130. "Overfishing: Plenty of Fish in the Sea? Not Always," *National Geographic* (as viewed February 12, 2015), http://ocean.nationalgeographic.com/ocean/critical-issues-overfishing/.

131. E.g., Andrew Tarantola, "The World's Largest Floating Fish Factory," *Gizmodo* (October 3, 2011), http://gizmodo.com/5845939/the-worlds-largest-floating-fish-factory.

132. Blue Water Fishermen's Association, "Pelagic Longlining" (as viewed February 13, 2015), http://www.bwfa-usa.org/our-fishery/pelagic-longlining.

133. Selina Haefeli, "Can Australia's Shores Cope with A Super-Trawler?" *Science Illustrated* (September 12, 2012), http://scienceillustrated.com.au/blog/features/can-australias-shores-cope-with-a-super-trawler/.

134. "Overfishing: Plenty of Fish in the Sea? Not Always."

135. Boris Worm et al., "Impacts of Biodiversity Loss on Ocean Ecosystem Services," *Science* 314 (3 November 2006): 787, 790.

136. Malin L. Pinsky et al., "Unexpected Patterns of Fisheries Collapse in the World's Oceans," *Proceedings of the National Academy of Sciences of the United States of America* 108 (May 17, 2011): 8317, doi:10.1073/pnas.1015313108.

137. The World Bank, "Fish to 2030: Prospects for Fisheries and Aquaculture," World Bank Report No. 83177-GLB, December 2013, 6, http://www.fao.org/docrep/019/i3640e/i3640e.pdf.

138. Ibid.

139. Ibid., 7 ("Given relatively stable capture fisheries in the last decades and the fact that dynamic biological processes determine the amount of fish stock available for harvest, modeling of price-responsive capture supply in a static sense seems unrealistic").

140. UN Food and Agriculture Organization, "The State of World Fisheries and Aquaculture" (2016), 2 and 3 fig. 1, http://www.fao.org/3/a-i5555e.pdf.

141. Ibid., 5.

142. Ibid., 2.

143. Matt McGrath, "Global Decline of Wildlife Linked to Child Slavery," *BBC News Science and Environment* (July 24, 2014), http://www.bbc.com/news/science-environment-28463036.

144. Ibid.

145. 2014 IPCC Synthesis Report, 67.

146. United Nations Convention on the Law of the Sea III, art. 61(3).

147. Ibid., art. 62(1), (2).

148. See generally, e.g., Rudiger Wolfrum and Nele Matz, "The Interplay of the United Nations Convention on the Law of the Sea and the Convention on Biological Diversity," in J. A. Frowein and R. Wolfrum, eds., *Max Planck Yearbook of United Nations Law* (Netherlands: Kluwer Law International, 2000), 445–480.

149. See 16 U.S.C. § 1856(a)(2) (extending state authority to the limits of the United States' territorial sea under the 1958 Geneva Convention on the Territorial Sea and Contiguous Zone, which was three nautical miles).

150. Ibid., § 1852(a).

151. Ibid., § 1852(h)(1).

152. Ibid., § 1854(a)–(c).

153. Ibid., § 1851(a).

154. Ibid., § 1851(a)(1).

155. Ibid., § 1802(33)(C).

156. Ibid., § 1802(34).

157. H.R. Rep. No. 94-445, 1976 U.S.C.C.A.N. 593, 614–615 (August 20, 1975).

158. 50 C.F.R. § 600.310(e)(i)(A).

159. UN Food and Agriculture Organization, "The State of World Fisheries and Aquaculture" (2014), 38, http://www.fao.org/3/a-i3720e.pdf ("Among the seven principal tuna species, *one-third of the stocks were estimated as fished at biologically unsustainable levels,* while 66.7 percent were fished within biologically sustainable levels [fully fished or underfished] in 2011. . . . *Market demand for tuna is still high and the significant overcapacity of tuna fishing fleets remains.*").

160. Ibid., 41.

161. 50 C.F.R. § 600.310(e)(1)(i)(A).

162. H.R. Rep. No. 94-445, 1976 U.S.C.C.A.N. 593, 615 (August 20, 1975).

163. 50 C.F.R. § 600.310(e)(1)(i)(A), (iv).

164. John H. Barnhill, "Maximum Sustainable Yield," in S. George Philander, ed., *Encyclopedia of Global Warming and Climate Change* (Thousand Oaks, CA: Sage Reference, 2012), 899–901.

165. R. Ian Perry, "Dealing with Uncertainty—Implications for Fisheries Adaptation," in Organisation for Economic Cooperation and Development (OECD), *The Economics of Adapting Fisheries to Climate Change* (2010), 149–158, doi:10.1787/9789264090415-en [hereinafter *OECD Fisheries Economics*]; see also Rögnvaldur Hannesson, "Climate Change, Adaptation, and the Fisheries Sector," in *OECD Fisheries Economics,* 247–275.

166. Edward Miles, "Fisheries Management and Governance Challenges in a Changing Climate," in *OECD Fisheries Economics,* 159–165.

167. Hannesson, "Climate Change, Adaptation, and the Fisheries Sector," 250–252.

168. Ibid., 262.

169. Miles, "Fisheries Management," 171.

170. Ibid., 167–171.

171. Ibid., 168.

172. Magnuson-Stevens Fishery Conservation and Management Reauthorization Act of 2006, Pub. L. No. 109-479, § 104, 120 Stat. 3575 (January 12, 2007); see also 16 U.S.C. § 1801(a)(6) (stating that the MSA was "necessary to prevent overfishing, to rebuild overfished stocks, to insure conservation, to facilitate long-term protection of essential fish habitats, and to realize the full potential of the Nation's fishery resources").

173. National Oceanic and Atmospheric Administration, NOAA Fisheries, US Department of Commerce, "Status of Stocks 2013: Annual Report to Congress on the Status of U.S. Fisheries" (April 29, 2014), 2, http://www.nmfs.noaa.gov/sfa/fisheries_eco/status_of_fisheries/archive/2013/status_of_stocks_2013_web.pdf.

174. 518 F. Supp. 2d 62 (D.D.C. 2007).

175. Ibid., 85 (citations omitted).

176. National Oceanic and Atmospheric Administration, NOAA Fisheries, "Status of Fish Stocks 2015" (2016): 1–2, http://www.nmfs.noaa.gov/sfa/fisheries_eco/status_of_fisheries/archive/2015/2015_status_of_stocks_updated.pdf.

177. Miles, "Fisheries Management," 167–168.

178. National Snow and Ice Data Center, "Climate Change in the Arctic," (as viewed February 16, 2015) https://nsidc.org/cryosphere/arctic-meteorology/climate_change.html. Specifically, "In the first half of 2010, air temperatures in the Arctic were 4° Celsius (7° Fahrenheit) warmer than the 1968 to 1996 reference period, according to NOAA." Ibid.

179. Ibid.

180. Craig Medred, "Expert Predicts Ice-free Arctic by 2020 as UN Releases Climate Report," *Arctic Dispatch-News* (November 2, 2014), http://www.adn.com /article/20141102/expert-predicts-ice-free-arctic-2020-un-releases-climate-report.

181. Chris Mooney, "The Incredible Decline of Arctic Sea Ice—Visualized," *The Washington Post,* April 3, 2015, https://www.washingtonpost.com/news/en ergy-environment/wp/2015/04/03/the-incredible-decline-of-arctic-sea-ice-visual ized/?utm_term=.5a6780ac7b51.

182. North Pacific Fishery Management Council, *Arctic Fishery Management* (as viewed May 28, 2017), https://www.npfmc.org/arctic-fishery-management/.

183. Ibid. For discussions of this closure, including a recommendation that this approach be incorporated as a climate change adaptation in fisheries management, see Sarah M. Kutil, "Scientific Certainty Thresholds in Fisheries Management: A Response to a Changing Climate," *Environmental Law* 41 (Winter 2011): 233; Jennifer Jeffers, "Climate Change and the Arctic: Adapting to Changes in Fisheries Stocks and Governance Regimes," *Ecology Law Quarterly* 37 (2010): 917.

184. US Department of State, "Arctic Nations Sign Declaration to Prevent Unregulated Fishing in the Central Arctic Ocean" (July 16, 2015), http://www.state .gov/r/pa/prs/ps/2015/07/244969.htm.

185. Ibid.

Chapter 6. Thinking Like a System: Resilience as a Narrative of Connection

Parts of this chapter were developed earlier in: Robin Kundis Craig, "What the Public Trust Doctrine Can Teach Us about the Police Power, *Penn Central*, and the Public Interest in Natural Resources: A Tribute to Joe Sax," *Environmental Law* 45 (2015): 519–559; Robin Kundis Craig, "Of Sea-Level Rise and Superstorms: The Public Health Police Power as a Means of Defending against 'Takings' Challenges to Coastal Regulation," *NYU Environmental Law Journal* 22 (2014): 84–115. Use of these previous works either conforms with the original copyright or is with permission of the publisher.

1. Aldo Leopold, *A Sand County Almanac and Other Writings on Conservation and Ecology,* Curt Meine, ed. (New York: Library of America, 2013). From the unpublished manuscript, "Wilderness ("The two great cultural advances . . . ")" (1935).

2. Aldo Leopold, *A Sand County Almanac, and Sketches Here and There* (New York: Oxford University Press, 1989), 204.

3. Daniel Bell, *Communitarianism and Its Critics* (Oxford: Oxford University Press, 2004 paperback edition), 1.

4. Amitai Etzioni, ed., *The Essential Communitarian Reader* (Lanham, MD: Rowman & Littlefield, 1998), 3.

5. Bell, *Communitarianism and Its Critics*, 14.

6. Etzioni, ed., *The Essential Communitarian Reader*, 3, xiii.

7. *Pennsylvania Coal v. Mahon*, 260 U.S. 393, 415 (1922).

8. Donald Worster, *Dust Bowl: The Southern Plains in the 1930s* (New York: Oxford University Press, 2004).

9. Aldo Leopold, "Engineering and Conservation," in *The River of the Mother of God and Other Essays by Aldo Leopold*, Susan L. Flader and J. Baird Callicott, eds. (1938; repr. Madison: University of Wisconsin Press, 1991), 249, 254.

10. Earth Systems Science is a close cousin to resilience theory and also reflects ideas reminiscent of the land ethic. Steven M. Stanley, *Earth System History* (London: Macmillan, 2005).

11. Leopold, *A Sand County Almanac*, viii–ix.

12. Ibid., 204.

13. Ibid., 132.

14. Eric T. Freyfogle. "Leopold's Last Talk," *Washington Journal of Environmental Law & Policy* 2 (2012): 236–281.

15. Ibid., 254–256.

16. Ibid., 241.

17. Ibid., 244.

18. Ibid.

19. Ibid., 249.

20. Ibid.

21. Worster, *Dust Bowl: The Southern Plains in the 1930s*, 6.

22. Leopold, *A Sand County Almanac*, viii.

23. Examples include the Animal Rights and Rights of Nature Movements. See Peter Singer, *Animal Liberation* (New York: HarperCollins, 2009); Colin Stone, *Should Trees Have Standing?* (New York: Oxford University Press: 2010).

24. Aldo Leopold received word that his manuscript had been accepted for publication by Oxford University Press on April 14, 1948, one week before he died of a heart attack while helping his neighbor fight a grass fire.

25. Leopold, *A Sand County Almanac*, 121.

26. Lennart Olsson, Anne Jerneck, Henrik Thoren, Johannes Persson, and David O'Byrne, "Why Resilience Is Unappealing to Social Science: Theoretical and Empirical Investigations of the Scientific Use of Resilience," *Science Advances* (2015): 1 of 11, doi:10.1126/sciadv.1400217.

27. Leopold, *A Sand County Almanac*, 205.

28. Katrina Brown, "Global Environmental Change I: A Social Turn for Resilience?," *Progress in Human Geography* (2013): 1–11, doi:10.1177/0309132513498837.

29. Millennium Ecosystem Assessment, 2005, *Ecosystems and Human Well-being: Current State and Trends* (Washington, DC: Island Press, 2005).

30. Ibid.

31. United Nations, "Millennium Development Goals and Beyond, 2015," http://www.un.org/millenniumgoals/. The goals are to eradicate extreme poverty and hunger; to achieve universal primary education; to promote gender equality and empower women; to reduce child mortality; to improve maternal health; to combat HIV/AIDS, malaria, and other diseases; to ensure environmental sustainability; and to develop a global partnership for development.

32. Leopold, *A Sand County Almanac*, 210.

33. 42 U.S.C. § 4332(C).

34. Melinda Harm Benson and Ahjond S. Garmestani, "Embracing Panarchy, Building Resilience and Integrating Adaptive Management through a Rebirth of the National Environmental Policy Act," *Journal of Environmental Management* 92 (2011): 1420–1427.

35. Melinda Harm Benson, "Intelligent Tinkering: The Endangered Species Act and Resilience," *Ecology and Society* 17 (2012): 28, http://dx.doi.org/10.5751/ES -05116-170428.

36. Frederic H. Wagner, "Whatever Happened to the National Biological Survey?," *BioScience* 49 (1999): 219–222, doi:10.2307/1313512.

37. Bruce Braun, "New Materialism and Neoliberal Natures," *Antipode* 47, no. 1 (2015): 1–14, 12, doi:10.1111/anti.12121. For more on this topic, see Sarah Nelson, "Beyond the Limits to Growth: Ecology and the Neoliberal Counterrevolution," *Antipode* 47, no. 2 (2014): 461–480, doi:10.1111/anti.12125.

38. David Upham, "The Primacy of Property Rights and the American Founding," *The Freeman* 98 (1998): 79–83.

39. U.S. Const., amend V (emphasis added).

40. Melinda Harm Benson, "The Tulare Case: Water Rights, the Endangered Species Act, and the Fifth Amendment," *Environmental Law* 32 (2002): 551–587. There is little in the way of historical evidence regarding why the founders included the compensation as a provision in the Fifth Amendment. Prior to the Bill of Rights, state governments often took property for roads and other public projects without compensating the owners; only the constitutions of Vermont and Massachusetts required that compensation be paid when private property was taken for public use. Ibid.

41. See Jorge L. Contreras, "A Brief History of Fraud: Analyzing Current Debates in Standard Setting and Antitrust through a Historical Lens," *Antitrust Law Journal* 80 (2015): 1, 52–53, 69–70 (discussing the Manufacturers Aircraft Alliance and the patent manipulations pushed by the US government during World War II in defense-related industries, including litigation resulting in a forced licensing). The US Supreme Court made clear in 2015 that patents are subject to the Takings Clause. *Horne v. U.S. Dept. of Agriculture*, ___ U.S. ___, 135 S. Ct. 2419, 2427-28 (2015).

42. *Loretto v. Teleprompter Manhattan CATV Corp.* (1982), 441.

43. Benson, "The Tulare Case."

44. *Village of Euclid, Ohio v. Ambler Realty Co.*, 272 U.S. 365, 366–367 (1926) (emphasis added).

45. In the United States, real property is often divided into two separate types of ownership—the mineral estate and the surface estate. The mineral estate includes not only coal and hard rock minerals but also oil, natural gas, and other fluid minerals below the surface. As a general legal matter, the mineral estate is the *dominant* estate in the sense that a mineral owner has the right to access the surface in order to remove the mineral resources.

46. *Pennsylvania Coal Co. v. Mahon* (1922), 415.

47. *Penn Central Transportation Company v. City of New York*, 438 U.S. 104 (1978).

48. Ibid., 117–118.

49. *Charles A. Pratt Construction Co. v. California Coastal Commission*, 162 Cal. App. 4th 1068, 1081–1083 (Cal. App. 2 Dist. 2008).

50. *U.S. Gypsum Co. v. Executive Office of Environmental Affairs*, 867 N.E.2d 764, 776–777 (Mass. Ct. App. 2007) (denying a landowner's taking claim in connection with restrictions on coastal development).

51. *Avenal v. Louisiana*, 886 So.2d 1085, 1101–1102 (La. 2004) (applying state public trust doctrine principles to help defeat a constitutional takings claim brought as a result of coastal restoration efforts); *Stevens v. City of Cannon Beach*, 854 P.2d 449, 454–457 (Or. 1993) (applying the state's doctrine of custom to defeat a takings claim when a coastal landowner was denied a permit to build a seawall for development).

52. 831 N.E.2d 865 (Mass. 2005).

53. Ibid., 867.

54. Ibid., 871 and n.13.

55. Ibid., 874 (emphasis added).

56. Ibid., 875.

57. *Borough of Harvey Cedars v. Karan*, 70 A.2d 524, 526 (N.J. 2013). The court also found that the land retained a value of $23,000 even with the building restriction. Ibid. Specifically, the court overruled the trial court's decision that, "in determining damages, . . . the jury [could not] consider that the dune would likely spare the Karans' home from total destruction in certain fierce storms and from other damage in lesser storms," ibid. Instead, in this partial takings case, the New Jersey Supreme Court determined that the jury had to consider the potential benefits—in terms of direct impacts on fair-market value—to the Karans' oceanfront property:

Harvey Cedars condemned a portion of the seaside, oceanfront property of the Karans to acquire a permanent easement for the construction and maintenance of a twenty-two-foot dune to replace an existing sixteen-foot dune. The new dune was part of a much larger shore-protection project to benefit all the residents of Harvey Cedars and Long Beach Island. Unquestionably, the benefits of the dune project extended not only to the Karans but also to their neighbors further from the shoreline. Yet, clearly the properties most vulnerable to dramatic ocean surges and larger storms are frontline properties, such as the Karans'. Therefore, the Karans benefitted to a greater degree than their westward neighbors. Without the dune, the probability of serious damage or destruction to the Karans' property increased dramatically over a thirty-year period.

A jury evidently concluded that the Karans' property decreased in value as a result of the loss of their panoramic view of the seashore due to the height of the dune. A willing purchaser of beachfront property would obviously value the view and proximity to the ocean. But it is also likely that a rational purchaser would place a value on a protective barrier that shielded his property from partial or total destruction. Whatever weight might be given that consideration, surely, it would be one part of the equation in determining fair market value.

Ibid., 541. See also Anne Siders, *Managed Coastal Retreat: A Legal Handbook on Shifting Development away from Vulnerable Areas,* Columbia University Climate Change Center (October 2013), iii ("Sales of coastal property should include a disclosure requirement that informs prospective purchasers of the risks they face.").

58. *In re: Property Located at 14255 53rd Ave., S., Tukwila, King County, Washington,* 86 P.3d 222, 223 (Wash. App. 2004), *review denied,* 103 P.3d 201 (Wash. 2004), *cert. denied sub nom Malbrain v. Washington State Dept. of Agriculture,* 544 U.S. 977 (2005) (quoting *Keystone Bituminous Coal Ass'n v. DeBenedictis,* 480 U.S. 470, 489 [1987]).

59. Marc R. Poirier, "The Virtue of Vagueness in Takings Doctrine," *Cardozo Law Review* 24 (2002): 93, 99.

60. Nicholas Blomley, "Performing Property, Making the World," *Canadian Journal of Law and Jurisprudence* 27 (2013): 23–48.

61. *A.M.L. International, Inc. v. Daley,* 107 F. Supp. 2d 90, 108 (D. Mass. 2000).

Conclusion. Living the New Story: Implications for Governance

Parts of this chapter were developed earlier in: Robin Kundis Craig, Ahjond S. Garmestani, Craig R. Allen, Craig Anthony (Tony) Arnold, Hannah Birgé, Daniel A. DeCaro, Alexander K. Fremier, Hannah Gosnell, and Edella Schlager, "Balancing Stability and Flexibility in Adaptive Governance: The New Challenges and a Review of Tools Available," *Ecology and Society* 22, no. 2, Special Issue, *Practicing*

Panarchy, online publication, 2017, article 3, http://www.ecologyandsociety.org/vol 22/iss2/art3/; Robin Kundis Craig, "Re-Tooling Marine Food Supply Resilience in a Climate Change Era: Some Needed Reforms," *Seattle University Law Review* 38 (2015): 1189–1235; Robin Kundis Craig and J. B. Ruhl, "Designing Administrative Law for Adaptive Management," *Vanderbilt Law Review* 67 (January 2014): 1–87; Robin Kundis Craig, "Climate Change Adaptation, the Clean Water Act, and Energy: A Call for Principled Flexibility Regarding 'Existing Uses,'" *George Washington Journal of Energy & Environmental Law* 4 (2013): 26–45; Robin Kundis Craig, "'Stationarity Is Dead'—Long Live Transformation: Five Principles for Climate Change Adaptation Law," *Harvard Environmental Law Review* 34 (2010): 9–73. Use of these previous works either conforms with the original copyright or is with permission of the publisher.

1. Greg Harman, "Your Brain on Climate Change: Why the Threat Produces Apathy, Not Action," *The Guardian,* November 10, 2014, https://www.theguardian .com/sustainable-business/2014/nov/10/brain-climate-change-science-psycholo gy-environment-elections.

2. Linus Blomqvist, Ted Nordhaus, and Michael Shellenberger, *Nature Unbound: Decoupling for Conservation* (September 2015), 12, http://thebreakthrough .org/images/pdfs/Nature_Unbound.pdf.

3. This quotation is widely attributed to Charles Darwin but may be apocryphal, akin to Mark Twain's supposed statement that "whiskey is for drinkin', and water is for fighting over." See John van Wyhe, "It Ain't Necessarily So . . . ," *The Guardian,* 8 February 2008, http://www.guardian.co.uk/science/2008/feb/09 /darwin.myths.

4. Amy Bentley, *Eating for Victory: Food Rationing and the Politics of Domesticity* (Champaign: University of Illinois Press, 1998), 4–5.

5. See generally Lewis Hyde, *Trickster Makes This World: Mischief, Myth, and Art* (New York: Farrar, Straus and Giroux, 2010).

6. Jellyfish are essentially the marine equivalent of cockroaches, surviving long after most other species die off. Robin Kundis Craig, "Avoiding Jellyfish Seas, or, What Do We Mean by 'Sustainable Oceans,' Anyway?," *Utah Environmental Law Review* 31 (2011): 17.

7. Roy Scranton, *Learning to Die in the Anthropocene: Reflections on the End of a Civilization,* (San Francisco: City Lights Open Media, 2015), 22.

8. Robin Kundis Craig, "'Stationarity Is Dead'—Long Live Transformation: Five Principles for Climate Change Adaptation Law," *Harvard Environmental Law Review* 34 (2010): 63–69, http://www.law.harvard.edu/students/orgs/elr/vol34_1/9 -74.pdf. For an example of how to use principled flexibility to change existing laws, see generally Robin Kundis Craig, "The Clean Water Act, Climate Change, and Energy Production: A Call for Principled Flexibility Regarding Existing Uses," *George Washington Journal of Energy & Environmental Law* 4 (2013): 27.

9. 16 U.S.C. § 1531.

10. U.S. Fish and Wildlife Service, "Greater Sage-Grouse," https://www.fws.gov/greatersagegrouse/ (last modified June 17, 2016).

11. Jessica B. Wilkinson, James M. McElfish, Jr., Rebecca Kihslinger, Robert Bendick, and Bruce A. McKenney, "The Next Generation of Mitigation: Linking Current and Future Mitigation Programs with State Wildlife Action Plans and Other State and Regional Plans," Environmental Law Institute and the Nature Conservancy (2009), http://www.eli.org/sites/default/files/eli-pubs/d19_08.pdf.

12. Ibid.

13. Robin Kundis Craig and J. B. Ruhl, "Designing Administrative Law for Adaptive Management," *Vanderbilt Law Review* 67 (January 2014): 1, 16–18.

14. Ibid., 27–40.

15. Ibid., 40–60, and Appendix (designing the Model Adaptive Management Procedure Act).

16. J. B. Ruhl, "Climate Change and the Endangered Species Act: Building Bridges to the No-Analog Future," *Boston University Law Review* 88 (2008), 1–62.

17. NOAA Fisheries, "TurtleWatch," Pacific Islands Fishery Science Center (as viewed September 17, 2016), https://pifsc-www.irc.noaa.gov/eod/turtlewatch.php.

18. J. Asafu-Adjaye, L. Blomqvist, S. Brand, B. W. Brook, R. Defries, E. Ellis, C. Foreman, D. Keith, M. Lewis, M. Lynas, T. Nordhaus, R. Pielke, R. Pritzker, J. Roy, M. Sagoff, M. Shellenberger, R. Stone, and P. Teague, "An Ecomodernist Manifesto," Ecomodernism.org (2015): 32, 11, http://www.ecomodernism.org/manifesto.

19. Ibid., 11.

20. Sandra L. Colby and Jennifer M. Ortman, "Projections of the Size and Composition of the U.S. Population: 2014 to 2060" (2015), 1-13, https://www.census.gov/content/dam/Census/library/publications/2015/demo/p25-1143.pdf.

21. "Use It and Lose It: The Outsize Effect of U.S. Consumption on the Environment," *Scientific American* (as viewed August 15, 2017), http://www.scientificamerican.com/article/american-consumption-habits/.

22. Robin Kundis Craig, Ahjond S. Garmestani, Craig R. Allen, Craig Anthony (Tony) Arnold, Hannah Birgé, Daniel A. DeCaro, Alexander K. Fremier, Hannah Gosnell, and Edella Schlager, "Balancing Stability and Flexibility in Adaptive Governance: The New Challenges and a Review of Tools Available," *Ecology & Society* 22, no. 2 (April 2017), article 3, http://www.ecologyandsociety.org/vol22/iss2/art3/ or https://doi.org/10.5751/ES-08983-220203.

23. J. B. Ruhl, "General Design Principles for Resilience and Adaptive Capacity in Legal Systems: Applications to Climate Change Adaptation Law," *North Carolina Law Review* 89 (2011): 1391, http://www.nclawreview.org/documents/89/5/ruhl.pdf.

24. Ibid.

25. Ibid, 1394.

26. Ibid, 1395.

27. Ibid, 1396. Dynamic federalism is an emerging challenge to traditional notions that the division of responsibilities across scales of governance promotes optimization and efficiency.

28. Ibid., 1398.

29. Craig & Ruhl, "Redesigning Administrative Law for Adaptive Management," 1.

30. Craig, "'Stationarity Is Dead,'" 63–66.

31. Ibid., 63–69.

32. Ibid., 40.

33. Melinda Harm Benson and Ahjond S. Garmestani, "Embracing Panarchy, Building Resilience and Integrating Adaptive Management through a Rebirth of the National Environmental Policy Act," *Journal of Environmental Management* 92 (2011): 1420 –1427.

34. The full quotation: "Reports that say that something hasn't happened are always interesting to me, because as we know, there are known knowns; there are things we know we know. We also know there are known unknowns; that is to say we know there are some things we do not know. But there are also unknown unknowns—the ones we don't know we don't know. And if one looks throughout the history of our country and other free countries, it is the latter category that tend to be the difficult ones." Defense.gov News Transcript: DoD News Briefing—Secretary Rumsfeld and Gen. Myers, United States Department of Defense (February 12, 2002), archive.defense.gov/transcripts/transcript.aspx?transcriptid2636.

35. Craig, "'Stationarity Is Dead,'" 43–44.

36. Ibid., 44.

37. Ibid., 46–47.

38. Ibid., 48–49.

39. Todd Woody, "Most Coal-Fired Power Plants in the US Are Nearing Retirement Age," *Quartz*, http://qz.com/61423/coal-fired-power-plants-near-retirement/ (March 12, 2013).

40. Craig, "'Stationarity Is Dead,'" 51–52.

41. Ibid., 55–58.

42. Ibid., 58–59.

43. G. D. Peterson, G. S. Cumming, and S. R. Carpenter, "Scenario Planning: a Tool for Conservation in an Uncertain World," *Conservation Biology* 17 (2003): 358–366, doi:10.1046/j.1523-1739.2003.01491.x.

44. Craig and Ruhl, "Redesigning Administrative Law for Adaptive Management," 1, 18–27 (making the same point).

45. Christine A. Klein, "The New Nuisance: An Antidote to Wetland Loss, Sprawl, and Global Warming," *Boston College Law Review* 48 (2007): 1155, 1158–1167.

46. Darren Botello-Samson, "Lawsuits, Property, and the Environment: Mea-

suring the Impact of Regulatory Takings Litigation on Surface Coal Mining Regulations" (August 31, 2006): 42–43 (unpublished manuscript), http://www.allaca demic.com/meta/p151975_index.html (suggesting that regulatory takings litigation can have a chilling effect on environmental and natural resources regulation).

47. For more in-depth discussions of these issues, see Robin Kundis Craig, "What the Public Trust Doctrine Can Teach Us about the Police Power, *Penn Central*, and the Public Interest in Natural Resources: A Tribute to Joe Sax," *Environmental Law* 45 (2015): 519–559; Robin Kundis Craig, "Public Trust and Public Necessity Defenses to Taking Liability for Sea-Level Rise Responses on the Gulf Coast," *Journal of Land Use & Environmental Law* 26 (2011): 395; Robin Kundis Craig, "Adapting to Climate Change: The Potential Role of State Common-Law Public Trust Doctrines," *Vermont Law Review* 34 (2010): 781; Robin Kundis Craig, "Adapting Water Law to Public Necessity: Reframing Climate Change Adaptation as Emergency Preparedness and Response," *Vermont Journal of Environmental Law* 11 (2010): 709.

48. See generally Robin Kundis Craig, "Of Sea-Level Rise and Superstorms: The Public Health Police Power as a Means of Defending against 'Takings' Challenges to Coastal Regulation," *New York University Environmental Law Journal* 22 (2014): 84–115.

49. 33 U.S.C. § 1313; 40 C.F.R. § 131.3 (2015).

50. J. B. Ruhl and Robert L. Fischman, "Adaptive Management in the Courts," *Minnesota Law Review* 95 (December 2010): 424, 426.

51. Scranton, *Learning to Die in the Anthropocene*.

52. Ibid., 19.

Index

hedging, 177
Heider, Fritz, 8
Heinlein, Robert A., 49
Heinrich, Martin, 100–101
herbicides, 27
Holling, C. S. "Buzz," 49, 61–63, 64
Holling school of resilience theory,
	57–60, 148. *See also* resilience
	theory
Holmes, Oliver Wendell, Jr., 153
human population growth, 172
"Humans as Controlling Engineers"
	narrative, 14–15
Hyde, Lewis, 50–51, 52, 195n15
hydroelectric dam development, 63

"If Antarctic Melting Has Passed the
	Point of No Return" (*Forbes*
	article), 19–20
incentive structures, 176
Industrial Revolution, 4–5, 160–161
instream flow water rights, 92
intentional transformation, 67
Intergovernmental Panel on Climate
	Change (IPCC)
	on changes to ocean currents, 118
	climate change and adaptation
		planning, 10
	on climate change and new realities
		of sudden, unexpected change,
		45
	on climate change and sustainable
		development, 41
	on climate change and threshold
		crossings, 9, 10
	climate change mitigation efforts
		and, 10
	on the failure of sustainability in the
		marine fisheries, 123–124
	on global sea-level rise, 119, 120
	on human influence on the climate
		system, 1
	impact of global climate change
		reports on policy efforts, 32

on the importance of
	transformational capacity, 74–75
on ocean acidification, 121
on predictability and reversibility of
	ecological change, 31
resilience theory and normative
	decision making, 69–70, 74
on rising ocean temperatures, 117
interlocking model of sustainable
	development, 38–39
International Commission on
	Stratigraphy (ICS), 4–5
International Union for the
	Conservation of Nature (IUCN),
	36, 38, 39, 40, 69–70, 74
International Union of Geological
	Sciences, 4
interstate compact agreements, 93
invasive species, 112–113
IPCC. *See* Intergovernmental Panel on
	Climate Change
"It Isn't Us" climate change narrative,
	13–14
"It's the End of the World as We Know
	It" climate change narrative,
	18–20
IUCN. *See* International Union for the
	Conservation of Nature

Jackson, Jeremy B. C., 116
Jemez River, 98–99
Jicarilla Apache, 90

"killer fogs," 4–5
King, Laura, 23
Kiribati, 108–109
Klum, Mattias, 42
Kohler Act (Pennsylvania), 153
Krakoff, Sarah, 33
Kuh, Katrina, 14

Lake Erie, 66
lake eutrophication, 65–66
land ethic, 136, 140–149, 158

impact of climate change and its
trickery on, 117–121
impact of climate change, synergy,
and complexity on, 121–124,
166–167
importance of applying resilience
theory to, 105–109
limitations of traditional narratives
of governance, 106–107
marine living resources concept,
105–106
panarchial interactions and, 104–105
pre–climate change human stressors,
111–117
resilience and coral reef ecosystems,
108–109
significance to humans, 105
uncertainties regarding the future of,
106–107
value of, 110–111
See also marine biodiversity; marine
fisheries
ocean acidification, 5, 16–17, 120–121
ocean currents, 118–119
ocean heat content (OHC), 117
ocean temperatures
coral bleaching and, 64
impact of climate change on,
117–118
impact on currents, 118–119
impact on marine fisheries, 122
sea-level rise and, 119–120
Old Man Coyote (trickster character),
54
on-channel surface water reservoirs, 93
open space preservation, 176–177
optimum yield, 125
Organisation for Economic Cooperation and Development
(OECD), 128
O'Sullivan, John Louis, 24
overfishing
defined in maximum sustainable
yield-based management, 125–126

impact on marine fisheries, 111,
115–117
oyster reefs, 116

Pacific Decadal Oscillation, 101, 104
Pacific Ocean
Great Pacific Garbage Patch,
114–115
panarchial interactions and, 104–105
resilience and coral reef ecosystems,
108–109
rising ocean temperatures, 117
Pacific sleeper sharks, 116
Pan (Greek god), 64
panarchy
concept of, 62–65
maximum sustainable yield-based
management and, 127
panarchial interactions in the
oceans, 104–105
trickster narrative and, 49
"Party Like It's 1999" climate change
narrative, 19–20
Pauly, Daniel, 122
*Penn Central Transportation Company v.
City of New York*, 153
Pennsylvania Coal Co. v. Mahon,
152–153
per se rule, 151, 154, 157
pesticides, 27
Phoenix Islands Protected Area, 108
physical takings doctrine, 151
phytoplankton, 105, 111, 122
planetary boundaries, 42–44
Planetary Boundaries Project, 42, 65,
169
Planet Earth, 27, 28
plastic pollution, 114–115
pollock fisheries, 116
pollution
effects of air pollution, 4–5
eliminating or reducing, 175–177
incentive structures, 176
See also marine pollution